ICIAM2023 Springer Series

Volume 12

This series aims to publish some of the most relevant results presented at the ICIAM 2023 held in Tokyo in August 2023.

The series is managed by an independent Editorial Board, and will include peer-reviewed content only, including the Invited Speakers volume as well as books resulting from mini-symposia and collateral workshops.

The series is aimed at providing useful reference material to academic and researchers at an international level.

Elena Braverman · Anatoliy Swishchuk
Editors

Delay and Stochastic Differential Equations

Modelling in Finance, Life Sciences, and Engineering

Springer

Editors
Elena Braverman
Department of Mathematics and Statistics
University of Calgary
Calgary, AB, Canada

Anatoliy Swishchuk
Department of Mathematics and Statistics
University of Calgary
Calgary, AB, Canada

ISSN 3091-3098 ISSN 3091-3101 (electronic)
ICIAM2023 Springer Series
ISBN 978-981-95-7309-7 ISBN 978-981-95-7310-3 (eBook)
https://doi.org/10.1007/978-981-95-7310-3

Mathematics Subject Classification: 34K12, 34K13, 34K27, 45J05, 65C30, 34F05, 34K06, 34K15, 34K20, 45M05, 60G66, 60H35, 91G10, 91G15, 34K25, 34K50, 45D05, 60F05, 60H10, 60G12, 60G55

This Springer imprint is published by the registered company Springer Nature Singapore Pte Ltd.
The registered company address is: 152 Beach Road, #21-01/04 Gateway East, Singapore 189721, Singapore

Preface

This volume presents the proceedings of lectures delivered at the *International Congress on Industrial and Applied Mathematics (ICIAM) 2023*, held in Tokyo, Japan, in August 2023. It features selected works from two mini-symposium sessions: "*Delay and Stochastic Differential Equations in Life Sciences and Engineering*" and "*Stochastic Modelling in Finance*." These sessions brought together leading researchers in delay and stochastic modeling, highlighting their diverse applications in finance, life sciences, and engineering.

The contributions compiled in this book represent a broad spectrum of theoretical developments and applied research in Delay and Stochastic Differential Equations (DDEs and SDEs). Together, they showcase the depth and versatility of current research in these fields and demonstrate how delay and stochastic effects serve as key mechanisms underlying complex dynamical behavior. Both delays and elements of uncertainty are ubiquitous in real-world applications, i.e., engineering, but there are two areas where delays and volatility cannot be ignored, and these are financial markets and processes in life sciences, such as DNA replications.

Each paper has undergone a rigorous peer-review process to ensure the highest standards of scholarship. These works collectively illustrate the richness of delay and stochasticity as drivers of nontrivial dynamical behavior, affirming the vitality of research at the interface of delay dynamics, stochastic modeling, and their numerical and applied extensions. From exact solutions of DDEs to adaptive schemes for SDEs, and from asymptotic analysis to financial applications, the diversity and depth of these contributions offer fertile ground for future exploration. Of particular interest is the recurring theme of the *constructive role of delay*—as a resonance mechanism, a source of periodicity, or a computational challenge—that continues to yield rich dynamical phenomena across disciplines.

The work by Kenta Ohira and Toru Ohira, "Delay and Resonance: From Differential Equations to Random Walks," stands out for its elegant demonstration of how delays can intrinsically generate resonance phenomena—without requiring external periodic forcing. The paper introduces three minimalistic models capturing frequency and amplitude resonance purely through delayed feedback. Of note is the analytical tractability of the first two delay differential models, including a rare exact solution expressible via Gaussian dynamics and another through Lambert's W function. The stochastic extension culminates in the concept of a "Delayed Random Walk," revealing resonance under suitable noise-delay configurations. This contribution underscores the role of delay as a constructive element in dynamic systems, enriching both deterministic and stochastic frameworks.

Complementing the Ohira and Ohira study, Ivanov et al.'s "Periodic Solutions of a Delay Differential Equation with a Periodic Multiplier" provides a rigorous exploration of periodic behaviors in scalar non-autonomous DDEs. Their construction of stable slowly oscillating periodic solutions—both with the fundamental and doubled period—emphasizes how periodicity in feedback and coefficients can be harnessed analytically. The results are particularly significant in systems biology and control theory, where delay and periodic forcing often interplay. Importantly, the authors validate the persistence of periodicity under smooth perturbations, lending robustness to their analytical findings.

Kelly and O'Donovan's "Adaptive Mesh Construction for the Numerical Solution of Stochastic Differential Equations with Markovian Switching" addresses a pressing computational challenge: numerically solving SDEs with Markovian switching. Their adaptive mesh method represents a meaningful advancement, particularly for stiff systems with state-dependent switching, such as those modeling cellular processes or network traffic. The method's convergence properties, validated both theoretically and empirically, highlight the benefits of hybrid schemes that balance efficiency and robustness. Their application to a biological SDE modeling telomere dynamics further demonstrates the approach's practical relevance.

The papers by Appleby and Lawless "Solution Space Characterisation of Perturbed Linear Functional and Integro-Differential Volterra Convolution Equations: Cesàro Limits" and "Characterisation of Asymptotic Behaviour of Perturbed Deterministic and Stochastic Pantograph Equations" delve into the subtleties of integrodifferential and pantograph equations under perturbations. Both works challenge common assumptions in classical theory, particularly regarding how perturbations influence long-term solution behavior. The discussion on Cesaro limits reveals surprising divergence–convergence relationships between perturbations and their solutions. Similarly, the pantograph study sharpens understanding of how deterministic and stochastic forcing alters asymptotic stability and decay rates. These results have implications for the theoretical underpinnings of stability theory of functional differential equations.

Two papers extend the discussion into quantitative finance. Swishchuk's "Stochastic Modelling and Applications of Big Data in Finance" offers a sweeping survey of point process models, including Hawkes and compound Hawkes processes, applied to high-frequency financial datasets. His broad modeling framework aligns with the realities of big data and algorithmic trading. In contrast, Cheng and Escobar-Anel's "Incomplete Market Analysis of Optimal Consumption and Robust Portfolio for the 4/2 Stochastic Volatility Model" focuses on a more specific topic of consumption-portfolio optimization under the 4/2 stochastic volatility model. The use of confluent hypergeometric functions to solve the associated utility maximization problem showcases both mathematical depth and practical ambition.

Calgary, Canada

Elena Braverman
Anatoliy Swishchuk

Contents

Delay and Resonance: From Differential Equations to Random Walks

Kenta Ohira and Toru Ohira

Abstract Resonance phenomena induced by delays are presented. We propose and discuss three very simple models. Unlike typical resonance phenomena, they do not require oscillating external forces. Therefore, the delay-induced mechanism can be the simplest in generating resonance. The first model is described by a simple non-autonomous delay differential equation. Its most notable feature is that its exact solution can be written down as a sum of Gaussian dynamics. We believe this is the first instance that such solution is obtained for non-autonomous delay differential equation. This capability enables the analysis of its frequency resonance behaviors. The second model is given by a slightly extended non-autonomous delay differential equation, whose exact solution can also be derived using Lambert's W function. This model exhibits amplitude resonance, the properties of which can be analytically understood. The third model is a delay stochastic dynamics that incorporates both noise and delay. With suitably tuned values of the noise strength and the delay, it also exhibits frequency resonant behavior. This model is described by a stochastic delay dynamical map, or by a special case of a random walk with delay ("Delayed Random Walk"). In the latter description, we can again analyze resonant phenomena. These examples illustrate yet another rich phenomenon brought about by the element of delay.

Keywords Resonance · Delay · Stochasticity · Random walks · Exact solutions

K. Ohira (✉)
Graduate School of Informatics, Nagoya University, Nagoya, Japan
e-mail: ohira.kenta.n7@f.mail.nagoya-u.ac.jp

T. Ohira
Graduate School of Mathematics, Nagoya, Japan
e-mail: ohira@math.nagoya-u.ac.jp

E. Braverman and A. Swishchuk (eds.), *Delay and Stochastic Differential Equations*, ICIAM2023 Springer Series 12, https://doi.org/10.1007/978-981-95-7310-3_1

1 Introduction

The phenomenon of resonance is one of the fundamental phenomena, particularly in physics. Its applications are wide-ranging, utilized from our daily lives, such as in microwave heating, to cutting-edge scientific experiments involving high-energy particles. Resonance generally occurs when an external vibration signal or perturbation corresponds to the inherent oscillation frequency of the target system. One notable characteristic is that while resonance induces significant motion, it doesn't necessarily require a strong external signal if the frequency matches appropriately. This is exemplified by the scenario of swinging a pendulum, where timing is crucial for generating large oscillations without requiring excessive force.

Building upon this fundamental concept and phenomenon of resonance, several variations can be considered. Here, resonance in terms of amplitude, called "amplitude resonance", and phenomena where specific frequencies are emphasized, meaning specific rhythms are highlighted, are also examined as "frequency resonance". One indicator of frequency resonance is when significant peaks appear in the power spectrum, reaching maximum values.

Furthermore, in more recent years, there is a phenomenon known as stochastic resonance (Benzi et al. 1982; Longtin et al. 1991; Moss and Wisenfeld 1995; Wiesenfeld and Moss 1995; Collins et al. 1995; Bulsara and Gammaitoni 1996; Gammaitoni et al. 1998; Ohira and Sato 1999; Tsimring and Pikovsky 2001; Priplata et al. 2002) that has garnered research attention. This occurs when there is noise included in a portion of the external force, and by appropriately adjusting the strength of this noise, resonance phenomena can be induced. There are various objects and studies related to this phenomenon, but one representative example is when noise reveals the period of very weak external signals, equivalent to frequency resonance (Rose et al. 1967; Longtin et al. 1991)

On the other hand, the presence of "delay," especially in feedback, as the main theme of this book, is well-known for generating complex behavior from oscillations. In various fields such as biology, mathematics, economics, and engineering, understanding the effects of delays has become an important issue (Der Heiden 1979; Bellman and Cook 1963; Cabrera and Milton 2002; Hayes 1950; Insperger 2006; Küchler and Mensch 1992; Longtin and Milton 1989; Mackey and Glass 1977; Milton et al. 2009; Ohira and Yamane 2000; Smith 2010; Stepan 1989; Stepan and Insperger 2006; Szydlowski and Krawiec 2010). Even for simple systems, delays in feedback and/or interactions can induce oscillations and complex behaviors."Delay Differential Equations" (DDEs) are crucial tools in such research. A representative example is the Mackey–Glass equation c, which exhibits various types of dynamics, including chaos. Although our understanding of delay systems and DDEs has gradually improved (e.g., Taylor and Campbell 2007), solving DDEs to obtain solutions is considered very difficult.

Based on this background, in this chapter, we introduce our research on resonance induced by delay (Ohira 2022; Ohira and Ohira 2023a, b, 2024, 2025; Ohira 2024). Specifically, we present three examples. The first two models extend well-known delay differential equations to non-autonomous systems (Busenberg and Cooke 1984; Ming 1990; Ford and Verduyn Lunel 2002; Győri and Horváth 2017) by introducing time-dependent coefficients. In one case, frequency resonance is observed, while in the other, amplitude resonance appears. Both of these delay differential equations (DDEs) can be solved analytically. In the former, solutions can be constructed as a sum of Gaussian dynamics through Fourier transforms, while in the latter, solutions can be obtained using Lambert's W function.

As the third example, we introduce a model that incorporates both delay and noise. This model can be considered a system in which delay and noise coexist, described as a special case of delayed stochastic differential equations (Küchler and Mensch 1992) or delayed random walks (Ohira and Milton 1995; Ohira 1997; Ohira and Yamane 2000). Again, phenomena analogous to frequency resonance are observed, and analytical solutions for the corresponding resonance peaks can be derived (Ohira and Sato 1999).

A key feature of these models, unlike typical resonance phenomena, is that they do not require oscillatory external forces. As a result, the expressions remain remarkably simple, making these mechanisms among the simplest for generating resonance. These examples highlight the rich dynamical phenomena introduced by the presence of delay.

2 Resonating Delay Differential Equation: Frequency Resonance

We have recently invesitgated the following delay differential equation (Ohira 2022; Ohira and Ohira 2025; Ohira 2024):

$$\frac{dX(t)}{dt} + atX(t) = bX(t - \tau) \tag{1}$$

where $a \geq 0$, b, $\tau \geq 0$ are real parameters, with τ interpreted as the delay.

This equation falls into a category known as non-autonomous delay differential equations (non-autonomous DDEs), which have been studied in a broader context (Busenberg and Cooke 1984; Ming 1990; Ford and Verduyn Lunel 2002; Győri and Horváth 2017). This equation can be also viewed as a slight modification of the well-studied Hayes equation (Hayes 1950), with only the second term replaced by a linear function of time rather than a constant. However, its behavior is quite distinct from that of the Hayes equation.

Particularly, for Eq. (1), oscillatory transient dynamics appear and disappear as the delay increases while maintaining asymptotic stability for $X = 0$. Also, frequency resonance is observed with the power spectrum peak of the dynamical trajectory reaching its maximum height when the delay is suitably tuned.

We first note some properties of the special cases of Eq. (1).

(i) With the delay $\tau = 0$ and the initial condition $X(t = 0) = X_0$, the solution to the equation is as follows:

$$X(t) = X_0 e^{-\frac{1}{2}at^2 + bt}. \tag{2}$$

Thus, this solution exhibits a Gaussian shape with its peak at b/a. Furthermore, with $b = 0$, we also note that Eq. (1) represents the annihilation operator equation of the quantum simple harmonic oscillator, with t interpreted as a position rather than time (e.g., Sakurai 1985). It has the Gaussian ground state as its solution.
(ii) The case where $a = 0$ is a special case of the Hayes equation (Hayes 1950). This equation has been much studied and we know the following.

- It is known that the dynamics of $X(t)$ monotonically approach to the asymptotically stable origin $X = 0$ in the range of

$$0 > b > -1/e\tau. \tag{3}$$

  With $b < -1/e\tau$, the oscillatory dynamics begin to appear.
- Including the above, in the range of

$$0 > b > -\pi/2\tau, \tag{4}$$

  $X = 0$ is asymptotically stable.
- Hence, for $b < 0$, the critical delay τ_c for the loss of the stability of the origin $X = 0$ is

$$\tau_c = -\pi/2b. \tag{5}$$

  At this point, $X(t)$ has a stationary sinusoidal solution with constant amplitude with the angular frequency $\omega_c = |b|$. Equivalently, the critical period of the oscillation is

$$T_c = 2\pi/\omega_c = 4\tau_c. \tag{6}$$

(iii) However, with $a > 0$ with a finite initial function, the asymptotic stability of the origin is kept for all delays. For the finite delays, the solution of Eq. (1) typically shows oscillatory dynamics.

2.1 Solution Through the Fourier Transform

Let us first analyze Eq. (1) analytically. The most notable feature is that an exact solution of (1), recently obtained by one of the authors Ohira (2024), can be explicitly constructed. Typically, solving delay differential equations involves finding a solution $X(t)$ for $t \in [0, \infty)$, given an initial function $\phi(t)$ for $t \in [-\tau, 0]$. In most cases, such solutions cannot be obtained explicitly. As a result, the primary focus of DDE analysis is on asymptotic stability and qualitative behavior.

Here, we take an indirect approach by solving equation (1) through the Fourier transform. First, we demonstrate that (1) can be solved explicitly in the Fourier-transformed space. Then, we apply the inverse Fourier transform to construct the solution for $t \in (-\infty, \infty)$. It is found that the exact solution can be expressed as a sum of Gaussian functions with peaks located at different time points.

Let us now concretely find a solution for Eq. (1) in the entire range of $t \in (-\infty, -\infty)$ without specifying boundary conditions. We can show that with $a > 0$, the solution converges to 0 for $t \to \pm\infty$ for finite solutions. Fourier transforming Eq. (1) yields:

$$i\omega\hat{X}(\omega) + ia\frac{d\hat{X}(\omega)}{d\omega} = -b\hat{X}(\omega)e^{i\omega\tau} \tag{7}$$

where

$$\hat{X}(\omega) = \int_{-\infty}^{\infty} e^{i\omega t} X(t)dt \tag{8}$$

Equation (7) can be easily solved, and the solution is given by:

$$\hat{X}(\omega) = \mathcal{C}Exp[-\frac{1}{2a}\omega^2 + \frac{b}{\tau a}e^{i\omega\tau}] \tag{9}$$

where $\mathcal{C}$ is an integration constant. Thus, we can assert the following:

The Fourier-transforme of a finite solution of equation (1) with $a > 0$ in the range $t \in (-\infty, -\infty)$ is given by (9). This transformed solution is unique up to the scale factor $\mathcal{C}$.

This leads to the following:

The finite solution of Eq. (1) with $a > 0$ in the range $t \in (-\infty, +\infty)$ can be obtained as follows through the inverse Fourier transform:

$$X_f(t) = \frac{1}{2\pi}\int_{-\infty}^{\infty} e^{-i\omega t}\hat{X}(\omega)d\omega = \frac{\mathcal{C}}{2\pi}\int_{-\infty}^{\infty} Exp[-\frac{1}{2a}\omega^2 + \frac{b}{\tau a}e^{i\omega\tau} - i\omega t]d\omega$$
$$= \frac{\mathcal{C}}{2\pi}\int_{-\infty}^{\infty} Exp[-\frac{1}{2a}\omega^2 + \frac{b}{\tau a}\cos(\omega\tau)]\cos(\frac{b}{\tau a}\sin(\omega\tau) - \omega t)d\omega \tag{10}$$

Again, the solution is unique up to the scale factor $\mathcal{C}$. Furthermore, this solution is given by the following expression (see Ohira 2024 for details of the derivation):

$$X_f(t) = \mathcal{C}\sqrt{\frac{a}{2\pi}}\sum_{n=0}^{\infty}\frac{1}{n!}\left(\frac{b}{a\tau}\right)^n e^{-\frac{1}{2}a(t-n\tau)^2} \tag{11}$$

This represents a superposition of Gaussian functions centered at $t = n\tau$, with a height of

$$\mathcal{C}\sqrt{\frac{a}{2\pi}}\frac{1}{n!}\left(\frac{b}{a\tau}\right)^n.$$

We can easily verify that Eq. (11) satisfies Eq. (1) through direct substitution.

To the best of our knowledge, this is the first instance in which such an explicit solution has been obtained for non-autonomous DDEs in general. We emphasize that although we expect this solution to be the unique finite solution of (1), a rigorous mathematical proof has yet to be established (Fig. 1).

We also observe in passing that in the special case of $\tau \to \infty$ as

$$\lim_{\tau\to\infty} X_f(t) = C\sqrt{\frac{a}{2\pi}}Exp[-\frac{1}{2}at^2] \tag{12}$$

2.2 Approximation of Initial Function Problem

Here, we note that the exact solution (11) can be used as an approximation for the standard initial value problem in DDEs.

Returning to Eq. (1) with $a > 0$, we seek a real solution in the range $t \in [0, \infty)$, given the initial function $\phi(t)$ for $t \in [-\tau, 0]$. In general, Eq. (1) with an arbitrary initial function remains unsolvable. However, the findings in the previous subsection suggest the following:

If the initial function $\phi(t)$ is given by the solution $X_f(t)$ from Eq. (11) for $t \in [-\tau, 0]$, then the solution $X_s(t)$ of Eq. (1) is simply $X_f(t)$ for $t \in [0, \infty)$.

In other words, if a specific initial function $\phi(t)$ matches $X_f(t)$ for $t \in [-\tau, 0]$, then the resulting solution $X_s(t)$ for $t \in [0, \infty)$ is directly given by $X_f(t)$ over the same range.

The exact solution can also serve as an approximation when the initial function $\phi(t)$ is not significantly different from (11) over $t \in [-\tau, 0]$. For instance, if we choose a constant initial function $\phi(t) = X_0$ for $t \in [-\tau, 0]$, it provides a reasonable qualitative approximation within certain parameter ranges. However, we note that the amplitude may differ significantly, requiring finer adjustments of the integration constant in (11) to better match the given initial condition. Some examples are illustrated in Figs. 2 and 3.

2.3 Power Spectrum and Frequency Resonance

Let us analyze the behavior of frequency resonance appearing with Eq. (1). We use the power spectrum calculated from Eq. (8).

$$S(\omega) = |\hat{X}(\omega)|^2 = \hat{X}(\omega)\hat{X}^*(\omega) = C^2 Exp[-\frac{1}{a}\omega^2 + \frac{2b}{\tau a}\cos\omega\tau] \tag{13}$$

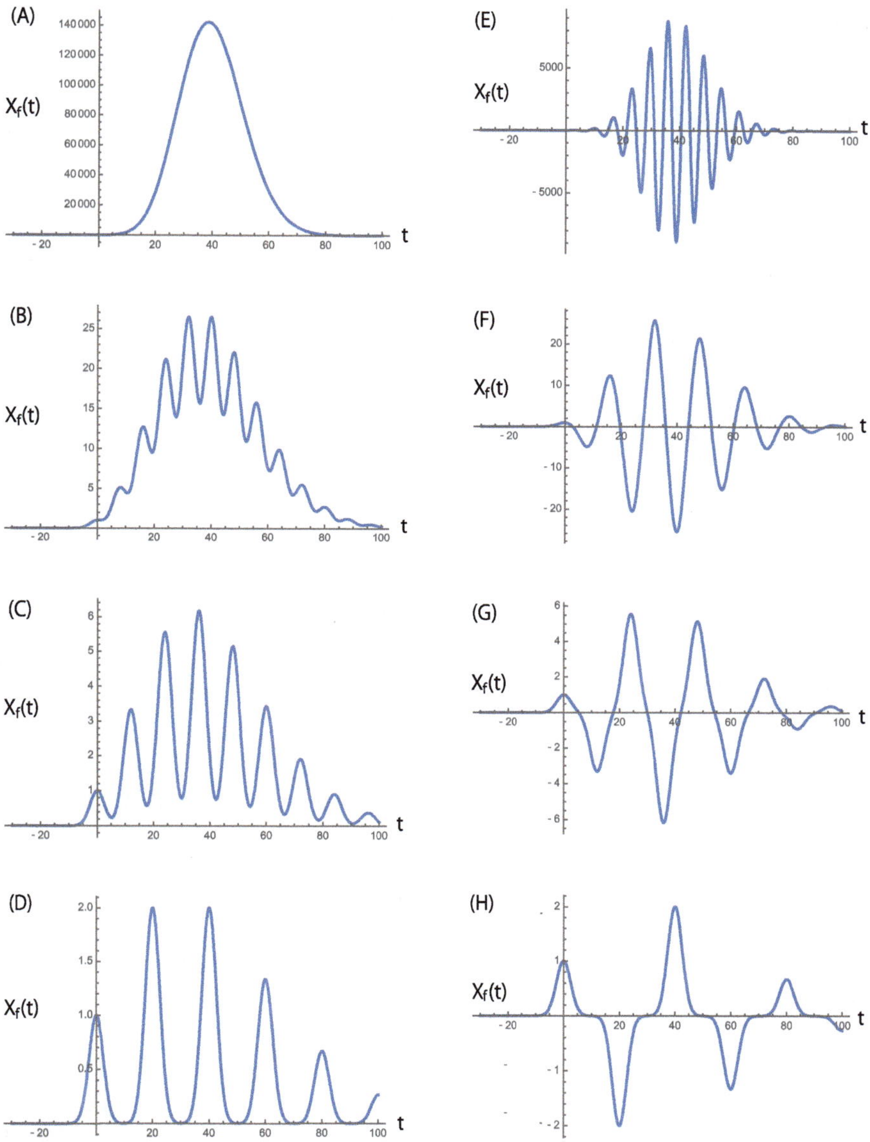

Fig. 1 Representative dynamics of equation (1) given by the solution (11). We set $\mathcal{C} = \sqrt{\frac{2\pi}{a}}$, $a = 0.15$, and sum the series up to $n = 500$. Other parameters (b, τ) are set as follows: **A** (6, 3), **B** (6, 8), **C** (6,12), **D** (6, 20), **E** (–6, 3), **F** (–6, 8), **G** (–6, 12), **H** (–6, 20)

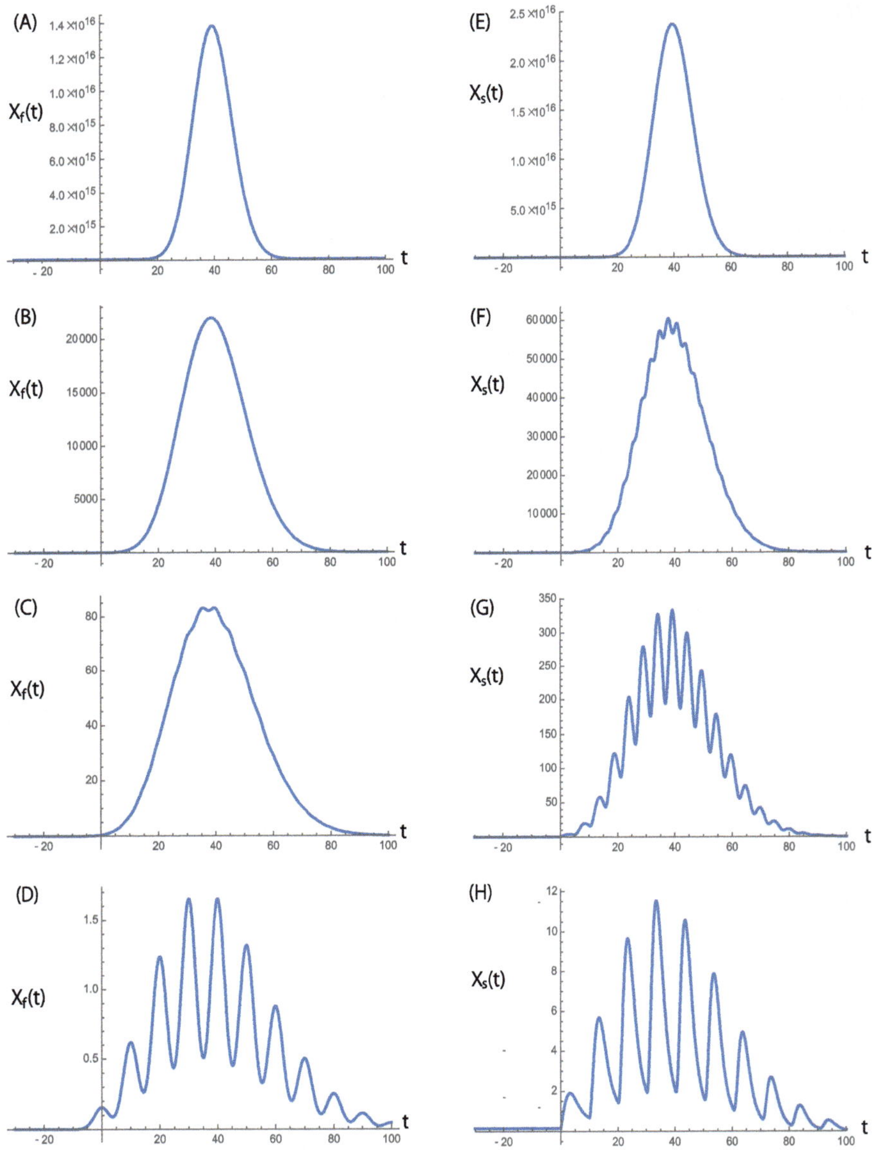

Fig. 2 Comparison of solutions $X_f(t)$ given by (11) (left column) and $X_s(t)$, numerical simulation dynamics of (1) (right column), with constant initial function set as $\phi(t) = X_f(0)$, $t \in [-\tau, 0]$. The parameters are set as follows; $a = 0.15$, $b = 6.0$, $C = 1$. The values of the delays τ are **A**, **E** 1, **B**, **F** 3, **C**, **G** 5, **D**, **H** 10

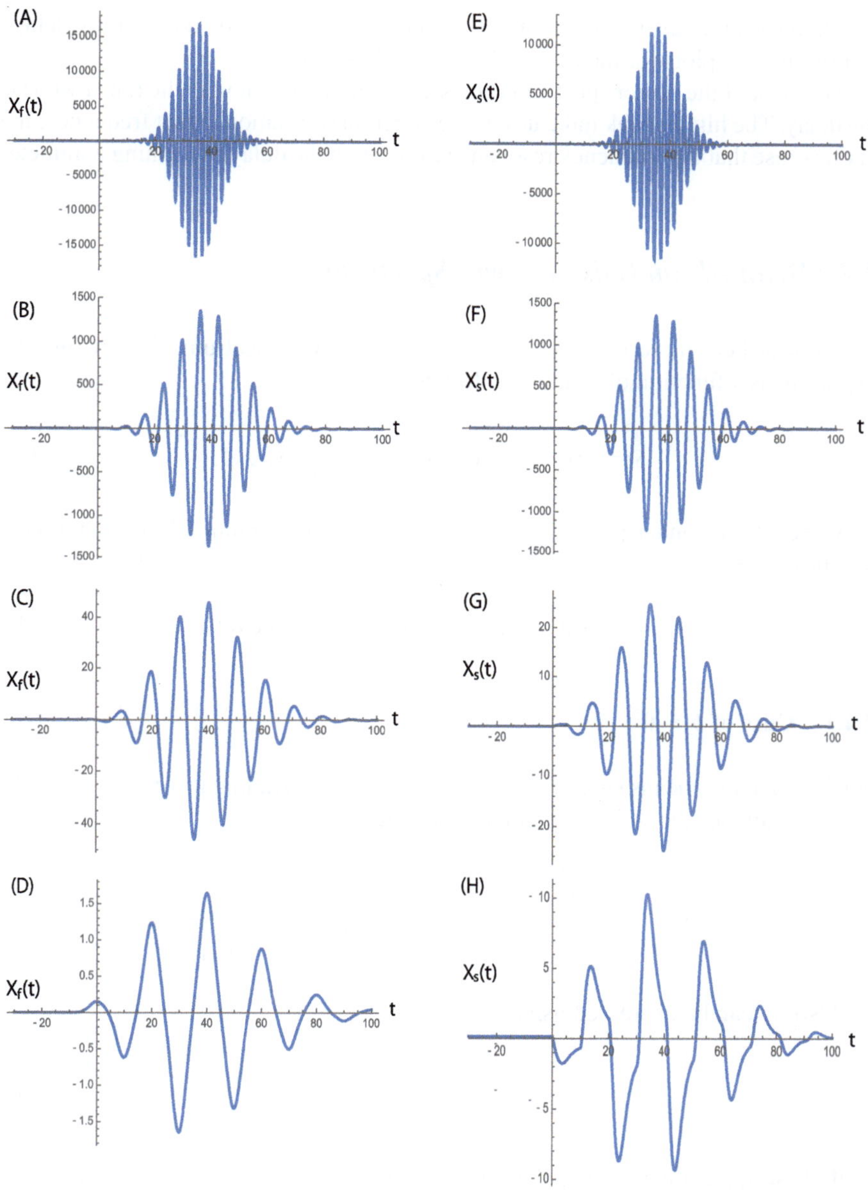

Fig. 3 Comparison of solutions $X_f(t)$ given by (11) (left column) and $X_s(t)$, numerical simulation dynamics of (1) (right column), with constant initial function set as $\phi(t) = X_f(0)$, $t \in [-\tau, 0]$. The parameters are set as follows; $a = 0.15$, $b = -6.0$, $C = 1$. The values of the delays τ are **A**, **E** 1, **B**, **F** 3, **C**, **G** 5, **D**, **H** 10

We have plotted this equation for the power spectrum for the various delays. Representative plots for the case of $b > 0$ are shown in Fig. 4.

The peak of the power spectrum shows a maximum height with the tuned value of the delay. The higher peak indicates, more coherent oscillation at that frequency. It is in this sense that the frequency resonance exists with the delay as a tuning parameter.

2.4 Detailed Analysis of Power Spectrum

We present here more thorough analysis of the power spectrum. We rewrite (13) explicitly as a function of ω and τ as follows:

$$S(\tau, \omega) = \mathcal{C}^2 Exp[-\frac{1}{a}\omega^2 + \frac{2b}{\tau a}\cos\omega\tau] \tag{14}$$

If we set the resonant point (τ_r, ω_r) that gives the maximum of this function, it satisfiesWe $\frac{\partial S}{\partial \omega}(\tau_r, \omega_r) = 0$ and $\frac{\partial S}{\partial \tau}(\tau_r, \omega_r) = 0$. Thus,

$$\omega_r = -b\sin\omega_r\tau_r, \quad \omega_r\tau_r\sin\omega_r\tau_r = -\cos\omega_r\tau_r. \tag{15}$$

2.4.1 Case of $b > 0$

For $b > 0$, these means $\sin\omega_r\tau_r < 0$ and $\cos\omega_r\tau_r > 0$, leading to $\frac{3}{2}\pi < \omega_r\tau_r < 2\pi$. Then,$\omega_r$ satisfies the following and uniquely determined.

$$\omega_r = -b\sin\sqrt{(\frac{b}{\omega_r})^2 - 1}, \quad \frac{b}{\sqrt{4\pi^2+1}} < \omega_r < \frac{b}{\sqrt{\frac{9}{4}\pi^2+1}}. \tag{16}$$

Also, τ_r can be expressed using ω_r:

$$\tau_r = \frac{\sqrt{b^2 - \omega_r^2}}{\omega_r^2} \tag{17}$$

With these τ_r, ω_r the power spectrum is at its maximum. Also, at this maximum the expression describing the power spectrum can be simplified by using (15) as follows:

$$Max(S) = S(\tau_r, \omega_r) = \mathcal{C}^2 Exp[\frac{\omega_r^2}{a}] \tag{18}$$

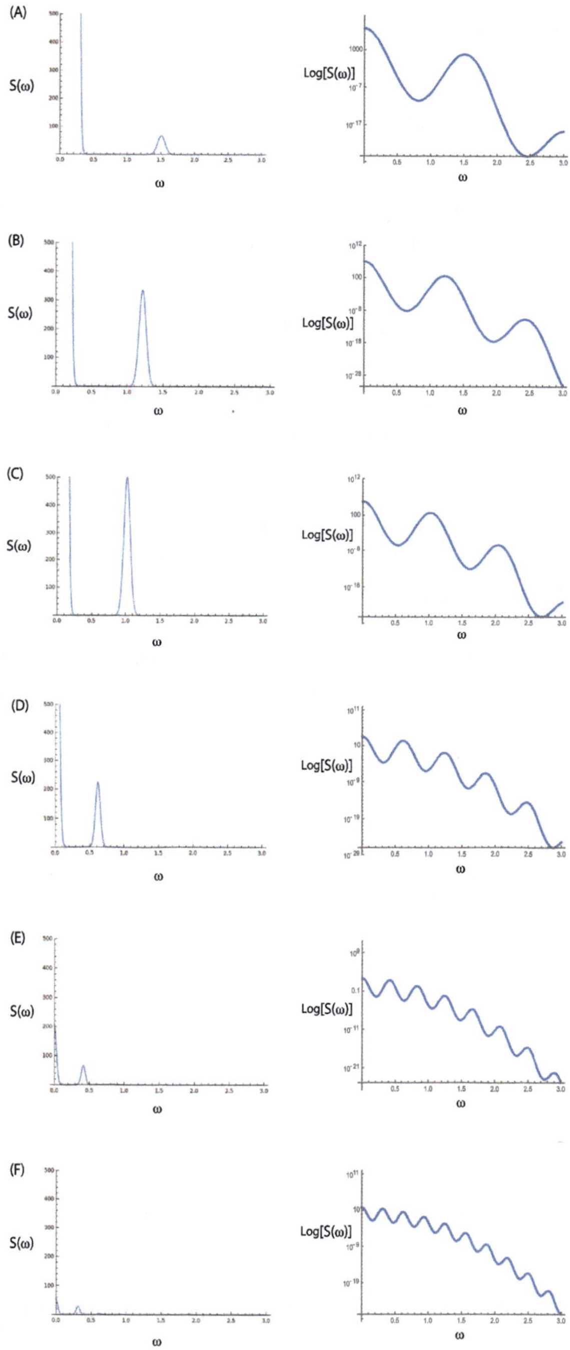

Fig. 4 Representative power spectrums given by the Eq. (13) with different values of the delays τ in linear (left column) and log (right column) scale. The parameters are set as follows; $a = 0.15$, $b = 6.0$, $C = 1$. The values of the delays τ are **A** 4, **B** 5, **C** 6, **D** 10, **E** 15, **F** 20

If we set $\omega_r = b\delta$ for (16), we obtain

$$\delta = -\sin\sqrt{(\frac{1}{\delta})^2 - 1}, \quad \frac{1}{\sqrt{4\pi^2+1}} < \delta < \frac{1}{\sqrt{\frac{9}{4}\pi^2+1}}. \tag{19}$$

By numerically solving above $\delta \approx 0.1612$ is attained, and using it for (17) and (18), the resonant point can be expressed as follows:

$$\tau_r = \frac{1}{b}\frac{\sqrt{1-\delta^2}}{\delta^2}, \quad \omega_r = b\delta, \quad Max(S) = S(\tau_r, \omega_r) = C^2 Exp[\frac{b^2}{a}\delta^2] \tag{20}$$

Further we can have $\tau_r \approx 37.9664/b$, $\omega_r \approx 0.1612b$ and $Max(S) \approx Exp[0.0259b^2/a]$ by using the numerical value of $\delta \approx 0.1612$

In order to confirm the above derivation, we examine Eq. (14) for the case with parameter values at $a = 0.15$ and $b = 6.0$. For these values, Eq. (20) gives $\tau_r \approx 6.327$, $\omega_r \approx 0.9673$, $Max(S) \approx 512.179$. We set the integration constant as $C = 1$. Comparing these with the numerical plot shown in Fig. 5A, the values of $\tau_r, \omega_r, Max(S)$ are consistent.
These results suggest that when $\tau = \tau_r$, oscillatory dynamics appears with the angular frequency $\omega = \omega_r$ as the major component.

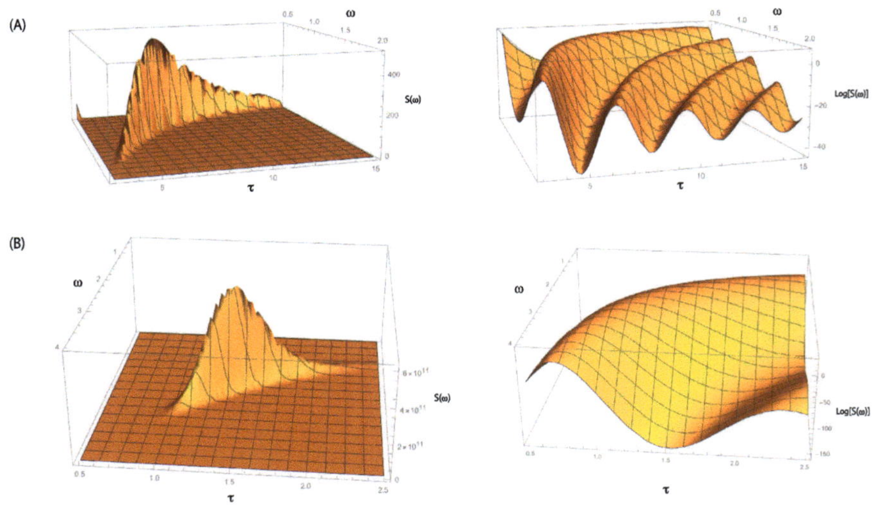

Fig. 5 3-D plot of the power spectrum as a function of ω and τ. The right column are plotted in the log scale. The parameters are set at $a = 0.15$, **A** $b = 6.0$, and **B** $b = -6.0$. The integration constant is $C = 1$

2.4.2 Case of $B < 0$

We can perform a similar analysis for the case $b < 0$ in Eq. (1). We observe similar transient dynamics along with power spectrum resonance. The representative dynamics and power spectra are shown in Fig. 5B.

Analyzing the power spectrum reveals that the resonance point (ω_r, τ_r) and the maximum of the power spectrum are given by:

$$\tau_r = -\frac{1}{b}\frac{\sqrt{1-\delta^2}}{\delta^2}, \quad \omega_r = -b\delta, \quad \max(S) = S(\tau_r, \omega_r) = C^2 \exp\left(\frac{b^2}{a}\delta^2\right), \quad (21)$$

where $\delta \approx 0.337$.

For $a = 0.15$ and $b = -6.0$, the maximum power $\max(S) = S(\tau_r, \omega_r) \approx 6.343 \times 10^{11}$ is attained at $(\tau_r, \omega_r) = (1.386, 2.019)$, which is again consistent with the numerical plot.

2.5 Summary

In this study, we proposed and analyzed a simple non-autonomous delay differential equation, denoted as Eq. (1). We demonstrated that its exact solution can be obtained using the Fourier transform over the entire real time axis. This led to an explicit solution expressed as a sum of Gaussian functions with different peak positions. Moreover, this solution can serve as an approximation for the standard initial function problem in DDEs for $t \in [0, \infty)$, provided that the prescribed initial function over $t \in [-\tau, 0]$ is not significantly different from the exact solution.

Using this approach, we were also able to analyze transient oscillatory behaviors exhibiting frequency resonances. In particular, we derived resonant conditions analytically.

In the analysis of delay differential equations, the predominant tools are often the Laplace transform or equivalent techniques, which are valuable for examining the stability characteristics of solutions. However, for Eq. (1), the Laplace transform does not yield an explicit solution, unlike the Fourier transform. The potential for further extending the Fourier transform approach to gain deeper insights into the transient dynamics of general delay differential equations remains an open avenue for future research.

3 Resonating Delay Differential Equation: Amplitude Resonance

Now, we turn our attention to the case of amplitude resoance and propose another delay differential equation, which is a slight extension of the equation proposed in the previous section.

$$\frac{dX(t)}{dt} + atX(t) = be^{-a\tau t}X(t-\tau). \tag{22}$$

With the exponent factor inserted in the right-hand side of the equation, it appears more complex to analyze. We, nevertheless, investigate this equation and show that previous analytical knowledge using the Lambert W functions can be employed.

3.1 *Analysis*

A recent analytical approach to simple DDEs uses the Lambert W function (Corless et al. 1996). It enables the expression of both formal and approximate solutions for the case with $a = 0$ of (22) (Shinozaki and Mori 2006; Pusenjak 2017; Ohira and Ohira 2023a).

$$\frac{dX(t)}{dt} = bX(t-\tau). \tag{23}$$

As discussed in the previous section, this is a special case of the Hayes equation (Hayes 1950).

The W function is defined as a multivalued complex function $W : C \to C$ satisfying

$$W(z)e^{W(z)} = z \tag{24}$$

The branches of the W function are expressed as $W_k, k = 0, \pm 1, \pm 2, \ldots, \pm\infty$. Using this function, the general solutions can be written as the following.

$$X(t) = \sum_{k=-\infty}^{\infty} C_k e^{\lambda_k t}, \quad \lambda_k = \frac{1}{\tau}W_k(b\tau). \tag{25}$$

We note that λ_k are the roots of the transcendental characteristic equation of (23),

$$\lambda = be^{-\tau\lambda}, \tag{26}$$

and that C_k are the constant coefficients determined by the initial interval condition $X(t) = \phi(t), [-\tau, 0]$ of (23).

The general solution of (22) can now be obtained by combining the above two cases. Namely, we set

$$X(t) = e^{-\frac{1}{2}at^2}\hat{X}(t), \tag{27}$$

then, it is easy to show that $\hat{X}(t)$ satisfies the same form of the differential equation as (23):

$$\frac{d\hat{X}(t)}{dt} = \hat{b}\hat{X}(t-\tau), \quad \hat{b} = \hat{b}(\tau) = be^{-\frac{1}{2}a\tau^2} \tag{28}$$

This leads to the general solution of (22) using (25) and (27) as

$$X(t) = e^{-\frac{1}{2}at^2}\sum_{k=-\infty}^{\infty} C_k e^{\frac{1}{\tau}W_k(\hat{b}\tau)t}. \tag{29}$$

We can infer qualitatively some properties of this solution.

- The first gaussian factor dominates as $t \to \infty$. Thus, for $a > 0$, the asymptotic stability of the origin is kept regardless of the value of the delay.
- The oscillatory behavior of the solution arises due to the second factor. Thus, as we have mentioned, the dynamics of $X(t)$ monotonically approach to the asymptotically stable origin $X = 0$ in the range of

$$0 > \hat{b}(\tau)\tau > -1/e. \tag{30}$$

With $\hat{b}(\tau)\tau < -1/e$, the oscillatory dynamics begin to appear.

In the next subsection, we first investigate the approximate dynamics from Eq. (22) using a formal solution (29).

3.2 Analysis

We utilize the formal form (29) of the solution of equation (22) in order to describe its dynamics. We argue that the formal solution (29) can be approximated for the ranges of parameters we are showing here by the following.

$$\begin{aligned} X_{W_0}(t) &= X_0 e^{-\frac{1}{2}at^2} Re[e^{\frac{1}{\tau}W_0(\hat{b}\tau)t}] \\ &= X_0 e^{-\frac{1}{2}at^2} e^{\frac{1}{\tau}Re[W_0(\hat{b}\tau)]t} \cos(\frac{1}{\tau}Im[W_0(\hat{b}\tau)]t) \end{aligned} \tag{31}$$

We first note that this solution is an exact solution of (22) over the range of $t \in (-\infty, +\infty)$ satisfying the condition that $X(t=0) = X_0$. Following the same argument as in the previous section, the solution $X_{W_0}(t)$ is the exact solution of (22), for $t > 0$ with the initial function given as $\phi(t) = X_{W_0}(t)$ for $-\tau \leq t \leq 0$. (We verify this through numerical comparison in the next subsection.)

As long as the initial condition is not too different, $X_{W_0}(t)$, that uses the single term of 0th (principal) branch of the W function instead of the sum over different branch terms, serves as a good approximation. The tabulated values of the W function (Appendix) are employed as well as some properties of the W function proved by Shinozaki and Mori (2006).

Let us consider the case that the initial function is given by a constant function $\phi(t) = X_0$ for $-\tau \leq t \leq 0$.

When x is a real value, the maximum value of the real part of the W function, $Re[W_k(x)]$, is obtained when

(i) $k = 0$, for $x > -1/e$
(ii) $k = 0$ and $k = -1$, for $x \leq -1/e$ with $Re[W_0(x)] = Re[W_{-1}(x)]$

In other words, the maximum values are the following for the real value x:

(i) $Re[W_0(x)]$, $(x > -1/e)$
(ii) $Re[W_0(x)] = Re[W_{-1}(x)]$, $(x \leq -1/e)$

Thus, the dominant term in the summation in (29) is either (i) $k = 0$ or (ii) $k = 0$ and -1 (see Appendix).

Another assumption we make is the following. The coefficients C_k in the sum of (29) need to be determined by the initial condition function $\{\phi(t) | -\tau \leq t \leq 0\}$. We, however, require the matching only at $t = 0$ for our case of constant initial condition function X_0.

Putting these rather physical assumptions and the above properties together we obtain the approximation (31) for the solution of (22) with initial constant function $\phi(t) = X_0$, $(-\tau \leq t \leq 0)$. Note that the case of (ii) with two terms $k = 0$ and -1 can also reduce to (31) by the properties that $Re[W_0(x)] = Re[W_{-1}(x)]$ and that $Im[W_0(x)] = -Im[W_{-1}(x)]$ which we have confirmed numerically in the Appendix.

The approximate solution also suggests transient oscillations due to the cosine factor. We can also use the W function to estimate the resonant point. From the structure of our approximation (31), we can infer that the maximum amplitude of the oscillation as we vary the delay can be obtained when the value of the exponent $g(\tau) = \frac{1}{\tau} Re[W_0(\hat{b}(\tau)\tau)]$ is positive and the largest. Thus, studying the function $g(\tau)$ can lead us to the optimal delay for the resonant point.

3.3 Comparison with Numerical Simulations

We are now in the position of studying Eq. (22) numerically for comparison against Eq. (31) and confirming some of the characteristics. As mentioned, the oscillatory behaviors appear and disappear as we increase the value of the delay τ, which can be considered a resonating phenomenon.

First, we verify that Eq. (31) is a exact solution of (22), for $-\tau <\leq$ with the initial function given as $\phi(t) = X_{W_0}(t)$ for $-\tau \leq t \leq 0$. The some results are shown in Fig. 6 where dots are plots of equation (31) using the tabulated numerical values for the W function, and the lines are direct numerical simulation of Eq. (22) with initial function of $\phi(t) = X_{W_0}(t)$. As we noted in the previous subsection, the asymptotic stability of the origin is kept even for large delays.

Now, we change the initial function to be a constant function $\phi(t) = X_0$, and numerically simulate Eq. (22). The typical dynamics for the case of $a > 0, b < 0$ are shown in Fig. 7. Comparing the plots of Figs. 6 and 7, the agreement with numerical results is fairly reasonable. Discrepancies become niticeable with larger delays (e.g. Figs. 6H and 7H) as more branch values $Re[W_k(\hat{b}(\tau)\tau)]$ of the W function become positive with increasing delays (see Appendix).

Also, we note that the amplitude of the oscillation changes and goes through the maximum as we increase the value of the delay (Fig. 8A) constructed from data of Fig. 7. In this sense, we have resonant phenomena with the delay as a tuning parameter. We can again use our solution (31) to estimate the resonant point at which the oscillation amplitude is maximal as we change the delay. As mentioned, this can be achieved by plotting the factor in the solution

$$g(\tau) = \frac{1}{\tau} Re[W_0(\hat{b}(\tau)\tau)] \tag{32}$$

and finding the positive maximum point numerically. It is given in Fig. 8B. The estimated resonant point is at $\tau \approx 2.13$, which agrees reasonably well with the one seen in Fig. 8A.

Thus, Eq. (22) is one of the simplest dynamical equations showing a resonance without any external oscillatory inputs. These properties are in contrast to the case of (23), where the stability of the origin is lost beyond the critical delay, and the amplitude of the oscillation is a monotonic increasing function of the delay.

3.4 Summary

Here, we have presented and investigated the delay differential equation (22) both analytically and numerically. Its behavior exhibits amplitude resonances. Furthermore, its solution is tractable, as it can be formally expressed in the form of Eq. (29) using the Lambert W function. We have demonstrated that even when utilizing only the principal branch of the W function, as shown in Eq. (31), transient dynamics can be approximated when the initial function is not significantly different. Previously, the W function has been employed for stability analysis (Shinozaki and Mori 2006) and for dynamical approximation in systems with fixed delays (Pusenjak 2017). However, our analysis extends its application to capturing dynamics with varying delays, which are typically considered challenging in the investigation of general delayed dynamical systems.

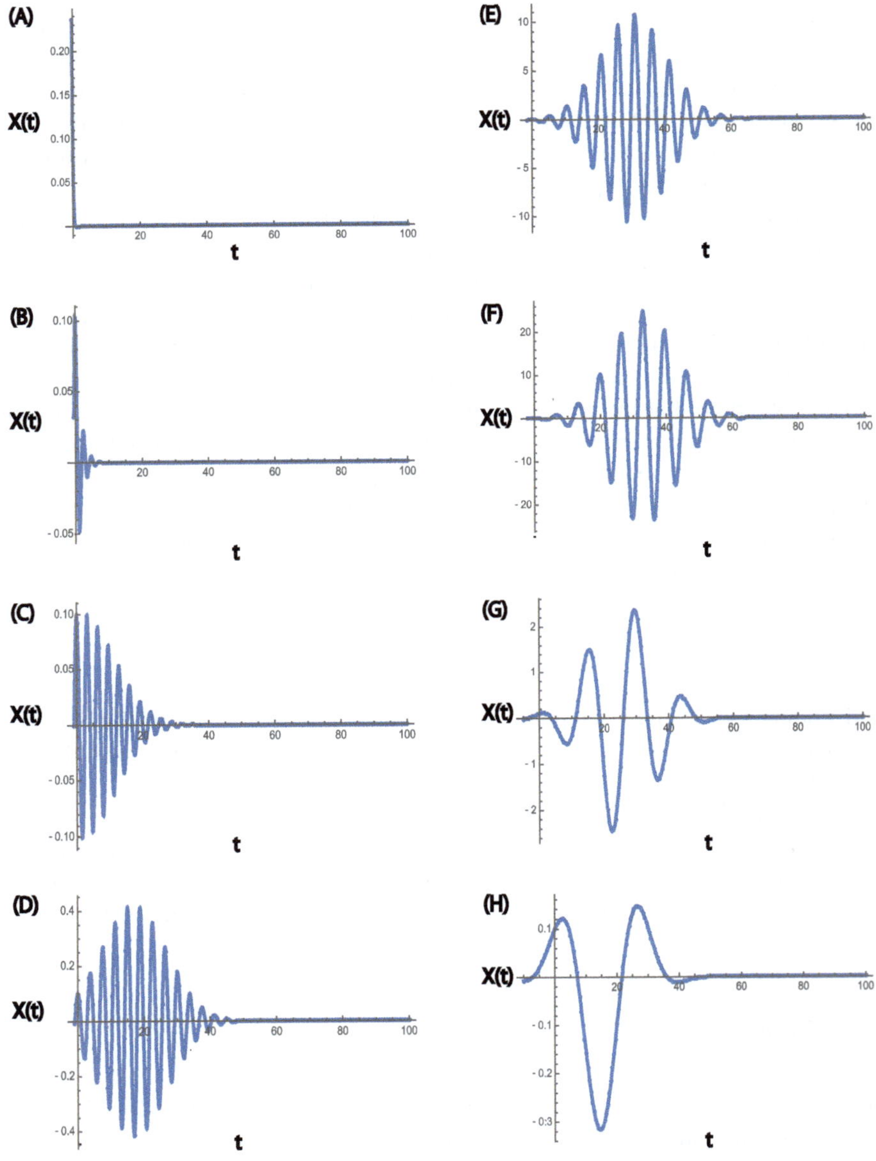

Fig. 6 Dots:Representative plots of the Eq. (31) with different values of the delays, τ Lines: Direct numerical simulation of (22) with the initial function given as $\phi(t) = X_{W_0}(t)$ for $-\tau \leq t \leq 0$. The parameters are set at $a = 0.01$, $b = -2.0$, $X_0 = 0.1$. The values of the delays τ are **A** 0.2, **B** 0.5, **C** 0.8, **D** 1.0, **E** 1.5, **F** 2.0, **G** 5, **H** 10

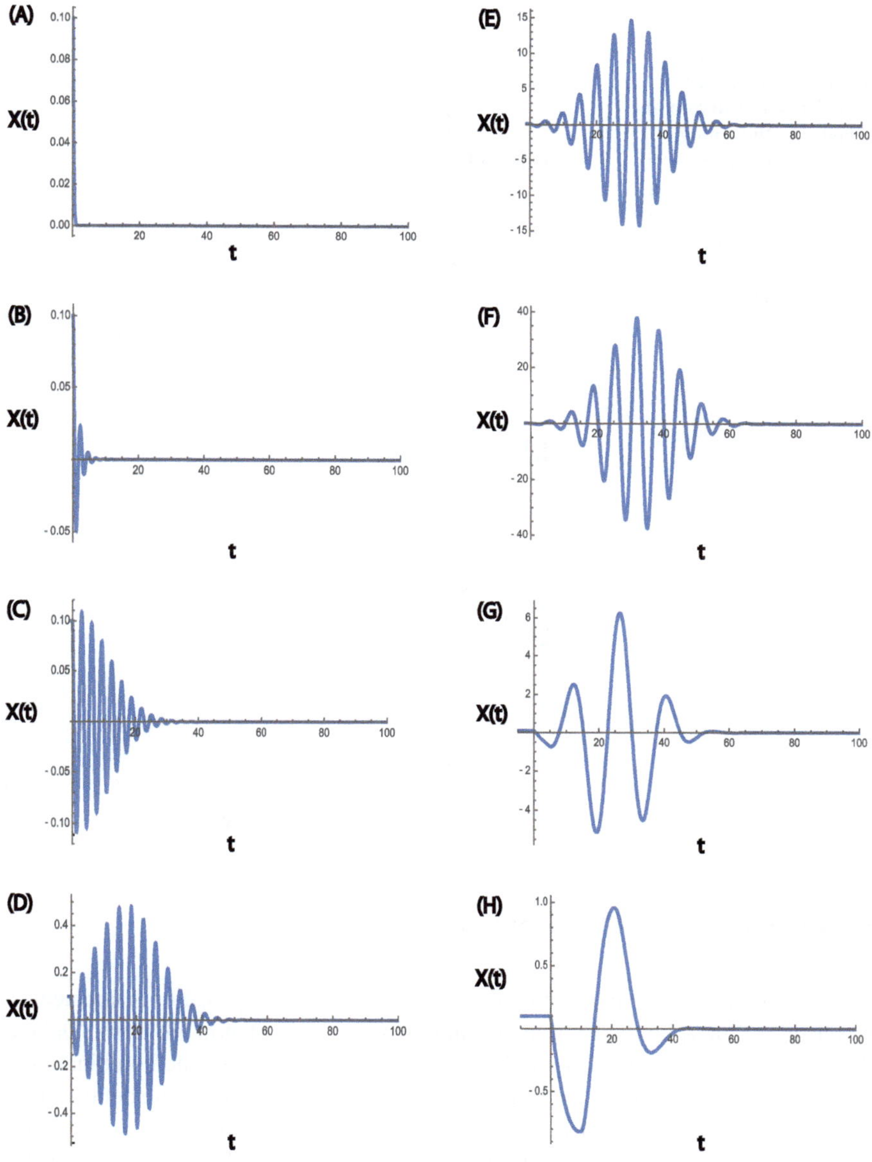

Fig. 7 Representative dynamics of equation (22) with the constant initial function as $\phi(t) = 0.1(-\tau \leq t \leq 0)$ for different values of the delays, τ. The parameters are set the same as in Fig. 8: $a = 0.01$, $b = -2.0$ The values of the delays τ are also the same: **A** 0.2, **B** 0.5, **C** 0.8, **D** 1.0, **E** 1.5, **F** 2.0, **G** 5, **H** 10

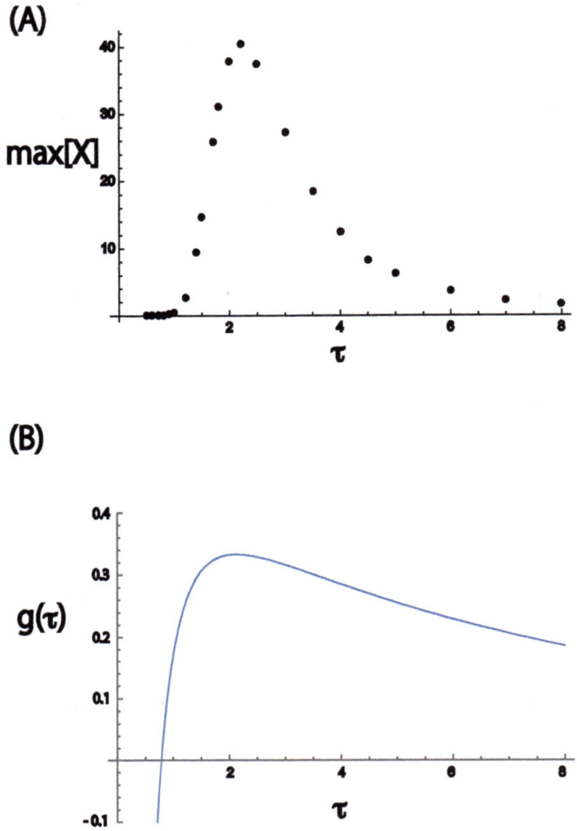

Fig. 8 **A** The resonant curve showing the maximum oscillatory amplitude, $max[X]$, with values of the delays τ as the tuning parameter. The parameters are set as the same as Fig. 1; $a = 0.01, b = -2.0$ with the initial interval condition as $X(t) = 0.1(-\tau \leq t \leq 0)$. **B** The plot of the function $g(\tau)$ given in (32) with $a = 0.01, b = -2.0$

4 Delayed Stochastic Resonance

We now proceed to the next topic, which explores the emergence of resonance as a result of the interplay between noise and delay, a phenomenon we term "Delayed Stochastic Resonance (Ohira and Sato 1999)." In this context, the combination of oscillatory behavior arising from stochasticity and that stemming from delay induces frequency resonance. We will examine this phenomenon from two perspectives: firstly, from the standpoint of stochastic dynamics, and secondly, from the perspective of a special case of random walks. Both descriptions will be provided.

4.1 Dynamical Map Model

We begin by introducing the dynamical model described by the following stochastic delay dynamical map:

$$X(t+1) = F[X(t-\tau)] + \xi_L = \frac{2}{1+e^{-\eta(X(t-\tau)-\theta)}} - 1 + \xi_L \tag{33}$$

Here η and θ are real parameters. The noise term, ξ_L, is a stochastic variable governed by the following probability distribution function:

$$P(\xi = u) = \begin{cases} \frac{1}{2L} & (-L \leq u \leq L), \\ 0 & (u < -L, u > L) \end{cases} \tag{34}$$

In this context, ξ_L represents a real random number generated within the range $(-L, L)$ with a uniform probability distribution. Furthermore, we assume that ξ_L possesses zero time correlations. We note that this model can be conceptualized as an dynamical map description of a simple stochastic neural cell voltage model featuring delayed self-feedback with synaptic weights represented by w (Ohira et al. 2000):

$$\mu \frac{d}{dt} V(t) = -V(t) + w\phi(V(t-\tau)) + \xi(t),$$

where:

$$\phi(V(t)) = \frac{2}{1+e^{-\eta(V(t)-\theta)}} - 1. \tag{35}$$

As we investigate the above dynamical map (33), it's worth noting that both time t and delay τ are restricted to integers. By appropriately tuning the parameters η and θ this map can be effectively considered as a dynamics in an asymmetric double well potential, as illustrated in Fig. 9A.

We conducted numerical simulations of this stochastic dynamical map (33) while varying the "noise strength" L, and the delay τ. Through careful adjustment of these parameters, a relatively regular "spiking pattern" emerges, as depicted in Fig. 9C. However, if the noise strength is either too weak or too strong for a given delay, this regular spiking pattern disappears, as shown in Fig. 9B, D.

These observations suggest the presence of the frequency resonance discussed earlier. However, analyzing the stochastic dynamics described by (33) proves to be challenging. This leads us to consider further simplification of these models, which will be discussed in the next subsection.

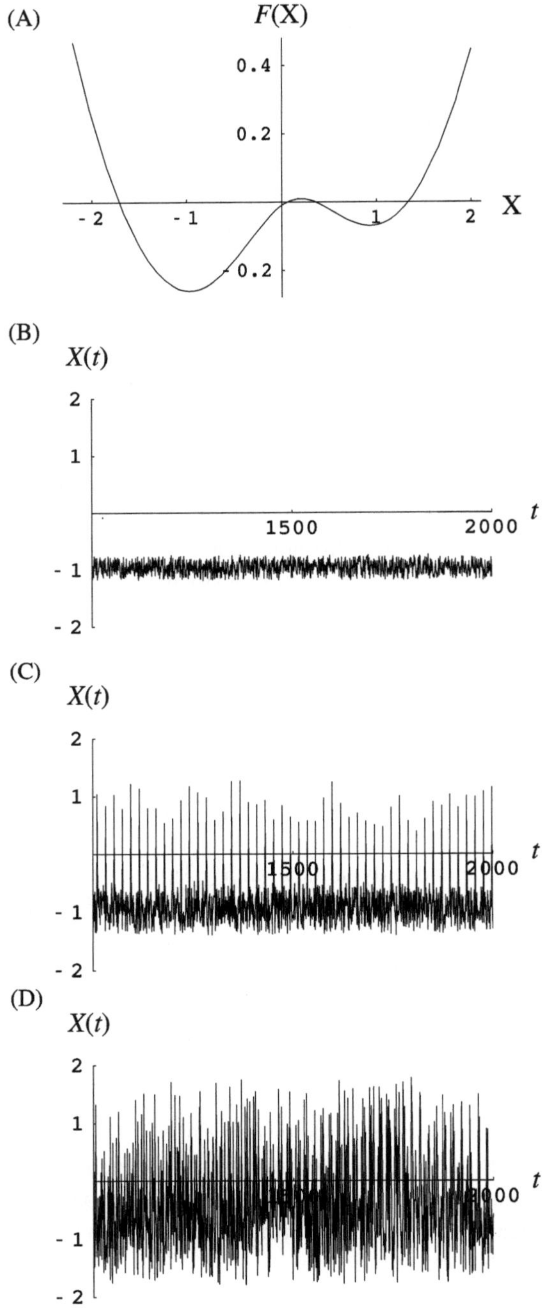

Fig. 9 **A** Asymmetric map function $\beta = 2.0$ and $\theta = 0.1$. **B–D** Representative dynamics of the stochastic map $X(t)$ as we change the noise width ("strength"), L. The values of L are **B** 0.2 **C** 0.4 **D** 0.8. Other parameters are set as $\tau = 20$, $\beta = 2.0$, $\theta = 0.1$, and the initial function for $t \in [-\tau, 0]$ is given by $X(t) = 0$. The plots are shown for the time step interval between $t = 1000$ and 2000

4.2 Stochastic Binary Bit

We now distill the essential features of the previous delayed stochastic dynamics and introduce the following delayed stochastic binary model. In this model, the state $X(t)$ assumes only two values, either ± 1 at time t. Utilizing the same random variable, ξ_L, we define the following two-state delayed stochastic dynamical map:

$$\begin{aligned} &X(t+1) = \theta[f(X(t-\tau)) + \xi_L], \\ &where: \\ &f(n) = \frac{1}{2}((a+b) + n(a-b)), \\ &and: \\ &\theta[n] = 1 \quad (0 \leq n), \qquad -1 \quad (0 > n) \end{aligned} \tag{36}$$

Here, a and b are real parameters satisfying $|a| \leq L$ and $|b| \leq L$ respectively. This model discretizes the dynamical map (33) into two states, and a and b control the probability of $X(t)$ taking a particular value based on its state at τ steps prior. By ensuring $a \neq b$, the probabilities for the two states of $X(t)$ are not symmetric, reflecting the asymmetry of the dynamical map (33). Furthermore, by setting:

$$p = \frac{1}{2}\left(1 + \frac{b}{L}\right), \quad q = \frac{1}{2}\left(1 - \frac{a}{L}\right) \tag{37}$$

we obtain a description in a probability space. In other words, as illustrated in Fig. 10A, the state of this stochastic binary bit at time $t+1$ is determined by its past state at the delay time interval τ. This can be expressed as:

$$P(1, t+1) = \begin{cases} p & X(t-\tau) = -1, \\ 1-q & X(t-\tau) = 1 \end{cases},$$

$$P(-1, t+1) = \begin{cases} q & X(t-\tau) = 1, \\ 1-p & X(t-\tau) = -1 \end{cases} \tag{38}$$

Here $P(s,t)$ represents the probability that $X(t) = s$. Thus, the probabilities of $X(t)$ assuming values of $+1$ or -1 are governed by its state in the past at $X(t-\tau)$. This model can be regarded as a special case of random walks with delay, termed "Delayed Random Walks" (Ohira and Milton 1995; Ohira 1997; Ohira and Yamane 2000).

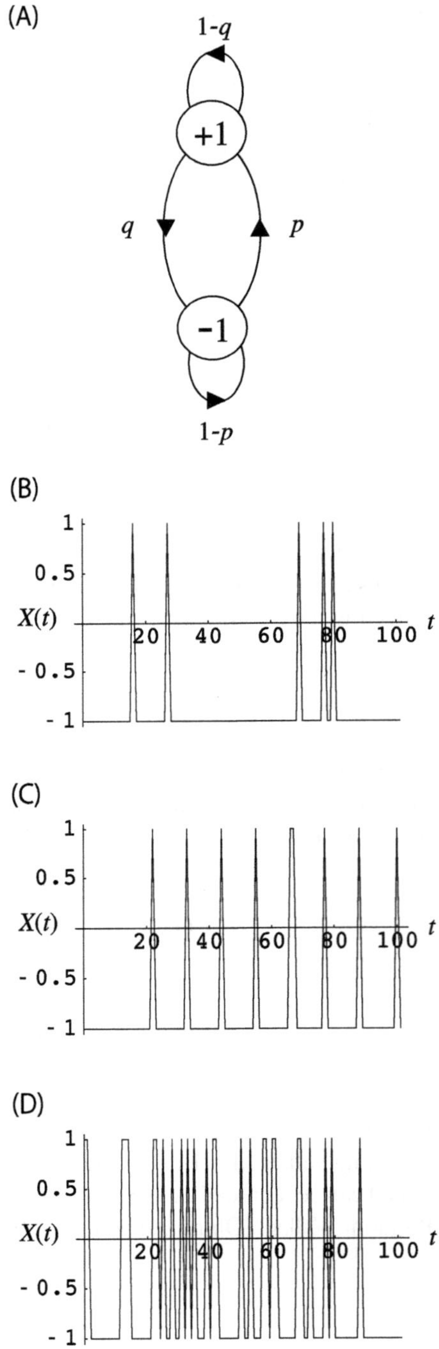

Fig. 10 **A** A schematic view of stochastic binary bit. **B–D** Representative dynamics of stochastic binary bit $X(t)$ as we change the parameter p. The values of p are set at **A** $p = 0.005$, **B** $p = 0.05$, **C** $p = 0.2$

4.3 Analysis and Numerical Simulation

We begin our investigation of the model with numerical simulations. To simplify, we set $q = 1 - q = 0.5$ and initialize the function for the interval of t in the range $[-\tau, 0]$, where the variable $X(t)$ takes values of ± 1 with equal probabilities. We vary the parameters p and τ, and record the value of $X(t)$ for up to 10^6 time steps. From this time series of $X(t)$, we construct a histogram, $h(u)$, representing the number of u consecutive time steps $X(t)$ takes the value of -1. Qualitatively, these histograms correspond to power spectrum.

In Fig. 10B–D, we present representative time series of $X(t)$, and the corresponding histgram $h(u)$ are shown in Fig. 11A. As $p < 0.5$, $X(t)$ typically takes the value of -1 more frequently and occasionally spikes to$+1$.

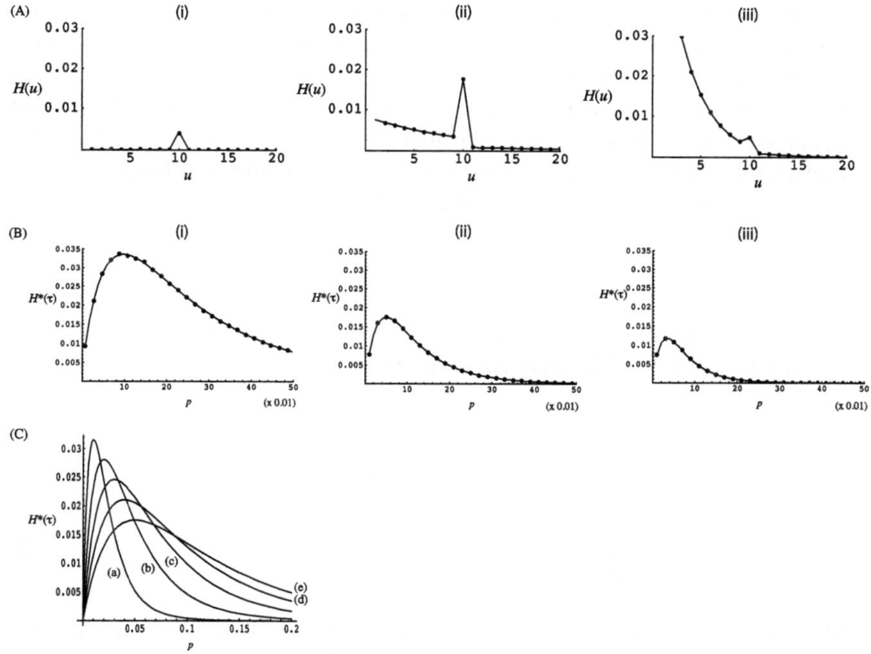

Fig. 11 **A** The histgrams for spike time intervals of dynamics of $X(t)$ as we change the parameter p. The values of p are set at (i) $p = 0.005$, (ii) $p = 0.05$, (iii) $p = 0.2$. Solid lines are from the mathematical analysis. **B** Plots of resoant peak heights as a function of p for different delays. The values of the delays are (i) $\tau = 5$, (ii) $\tau = 10$, (iii) $\tau = 15$. Solid lines are from the mathematical analysis. **C** Plots of resonant curve as we change the parameter q. The delay is set at $\tau = 10$ and the values of q are **a** 0.1, **b** 0.2, **c** 0.3, **d** 0.4, **e** 0.5

Specifically, for $p \ll 0.5$, the bit occasionally spikes to $X = +1$, as shown in Fig. 10B. However, these spiking behaviors are sparse and do not occur in series, resulting in small peaks in the corresponding histogram (Fig. 11A-i).

On the contrary, as p approaches 0.5, the state $X = +1$ appears more frequently. Nonetheless, regular spiking patterns, as seen in Fig. 10D, are not observed. The corresponding histogram shows peaks with relatively small heights accordingly (Fig. 11A-iii).

However, when the value of p is appropriately tuned given the value of τ, the time series exhibit more regular spiking behavior with time step intervals of approximately τ (Fig. 10C). Consequently, the corresponding histogram displays higher peaks, as depicted in Fig. 11A-ii.

The presence of maximum height in the peaks of the histogram and/or power spectrum indicates the existence of frequency resonance, as discussed in previous sections. Specifically, the frequency resonance emerges through the appropriate tuning of noise strength or probability parameters, termed as Stochastic Resonances as described earlier. Our simple model here can be regarded as a delay-induced stochastic resonance or a resonance involving noise and delay.

We now proceed to mathematically analyze this model to understand its behavior. The characteristic of this model is that given τ, it comprises $\tau + 1$ independent Markov chains, each appearing at intervals of $\tau + 1$. Utilizing this characteristic, we express t using two integers s and k as follows:

$$t = s(\tau + 1) + k \quad (0 \le s, 0 \le k \le \tau) \tag{39}$$

Equivalently, we can represent that $P_{\pm}(t) \equiv P_{\pm}(s, k)$. Utilizing the Markov properties, we can derive the following expressions:

$$\begin{aligned}
&P_{+}(s, k) = \alpha(1 - \gamma^{s}) + \gamma^{s} P_{+}(s = 0, k),\\
&P_{-}(s, k) = \beta(1 - \gamma^{s}) + \gamma^{s} P_{+}(s = 0, k)\\
&where:\\
&\alpha = \frac{p}{p+q}, \quad \beta = \frac{q}{p+q}, \quad \gamma = 1 - (p + q).
\end{aligned} \tag{40}$$

From the above equations, we observe that in the long-time limit of the stationary state, $P_{+}(s \to \infty, k) \equiv P_{+} = \alpha$ and $P_{-}(s \to \infty, k) \equiv P_{-} = \beta$, which are physically consistent.

We can also derive an expression for the histogram in the stationary state by defining $H(u) \equiv P(+; -, u; +)$. This represents the stationary probability for a bit to take $X = -1$ for u consecutive time steps between two $X = +1$ states. Utilizing this definition and the property that this model can be described by a series of independent Markov chains, we arrive at the following expressions:

$$P(+; -, u; +) \approx P_+(P_-)^u P_+ = (\beta)^u(\alpha)^2 \quad (1 \le u < \tau) \tag{41}$$

$$\approx P_+(P_-)^\tau(1-q) = (\beta)^\tau(\alpha)(1-q) \quad (u = \tau) \tag{42}$$

$$\approx P_+(P_-)^\tau(q)(1-p)^{u-\tau}(p) = (\beta)^u(p)^2 \quad (u > \tau) \tag{43}$$

By appropriately normalizing, this analytical expression accurately captures the numerical results, as shown in Fig. 11(A).

Notably, there is a distinct peak at $u = \tau$, which reaches its maximum as p varies. Additionally, the background—i.e., the values of $H(u)$ outside the peak—decays more rapidly as p increases, indicating a longer tail for smaller values of p.

Furthermore, we can obtain the resonance condition where the resonant peak becomes the highest. This can be achieved by taking the derivative of the expression (42) with respect to p and setting it to zero:

$$q \approx p\tau \tag{44}$$

We have plotted in Fig. 11B how the maximum value, H^*, changes as a function of p and τ.

The resonance condition dictates that these curves reach their maximum; when τ is fixed, adjusting p according to (44) allows us to achieve the strongest frequency resonance.

Although we won't delve further into this discussion, it's noteworthy that by controlling the value of q, as depicted in Fig. 11C, we can influence both the height and width of the resonant curves.

4.4 Summary

The delayed stochastic bit model, while simple, allows us to analyze associated delayed stochastic resonance mathematically. Most investigations of stochastic resonance rely on computational simulations and approximate theories. From this perspective, this model provides rare exact analytical insights.

There are extensions of these investigations. Firstly, a model incorporating both the current state $X(t)$ and the past state $X(t-\tau)$ in stochastic dynamics has been constructed and studied (Tsimring and Pikovsky 2001). The analysis is more complex, but multiple resonant peaks are still observed in the power spectrum.

Secondly, this model has been extended to networks of delayed stochastic bits (Ohira 2002). Again, frequency resonance is observed, although the analysis becomes more intricate.

Finally, the phenomena of delayed stochastic resonance we have suggested are confirmed by multiple experiments using lasers, indicating possible engineering applications (Massoller 2002; Misono et al. 2009).

These investigations suggest the wider possibility of applications of delayed stochastic resonances.

5 Discussion

In this article, we reviewed our investigation concerning resonances induced by delays. As mentioned, these proposed models can generate resonant behaviors without explicit oscillating external inputs. Another key point is that these models and their associated resonating behaviors are analytically tractable in one way or another. Thus, we provided rare "solvable" delay systems, allowing insights not only into asymptotic stabilities but also transient behaviors. We hope that these models can contribute to a deeper understanding of nonlinear transient phenomena induced by delays.

We make several comments for future directions:

1. We can consider equations of a more general form of (1), given as:

$$\frac{dX(t)}{dt} + atX(t) = f(X(t-\tau)), \tag{45}$$

   where the function f can be nonlinear. Our preliminary simulation results show that by setting f in the form of the "negative feedback function" and the "mixed feedback function" (Mackey and Glass 1977; Glass et al. 1988; Glass and Mackey 1988), we can observe transient oscillations and chaotic dynamics respectively. They are also suppressed with asymptotically larger delay. However, further investigations are needed to understand the nature of these behaviors with the general equation.
2. Our proposed models and behaviors are different from delay-induced transient oscillation (DITO) (Milton et al. 2010; Pakdaman et al. 1998). DITO is a solution of a coupled set of delay differential equations. With DITO, the transient oscillations are not suppressed as in our model, but oscillatory behaviors maintain a prolonged duration of oscillation with increasing delay. Whether our or similar analytical approaches can be used for DITO is open for future studies.
3. Other than reviving dynamics, we can consider various non-autonomous delay differential equations that can be described by generalizing Eq. (22) (Ohira and Ohira 2024). Exploration of some other peculiar dynamics induced by delays is left for the future.

4. Another intriguing observation is that Eq. (1) with $b = 0$ represents the annihilation operator equation for Quantum harmonic oscillators. Further differentiation of (1) with respect to time yields a second-order delay differential equation. Exploring this direction and potential connections to Quantum mechanics is left for future investigations.
5. There is much to be studied for the resonance with delays and noise. For example, we can consider a variety of different types of time-correlated noise. Correlation time and delay together may produce other peculiar behaviors. Together with the experimental confirmation and applications of this phenomena, such theoretical extensions can pose interesting and challenging research topics.

Competing Interests The authors have no conflicts of interest to declare that are relevant to the content of this chapter.

Acknowledgements The authors extend their gratitude for the valuable discussions with Prof. Hideki Ohira and the members of his research group at Nagoya University. This work was supported by the "Yocho-gaku" Project sponsored by Toyota Motor Corporation, JSPS Topic-Setting Program to Advance Cutting-Edge Humanities and Social Sciences Research (Grant Number JPJS00122674991), JSPS KAKENHI (Grant Number 19H01201), Ohagi Hospital (Hashimoto, Japan), and the Research Institute for Mathematical Sciences, an International Joint Usage/Research Center located at Kyoto University.

Appendix

In this appendix, we tabulate the values of the W function that are used in and related to the main text of this paper (Tables 1 and 2). The values are from Mathematica(R), ver. 13. We also plot them in Figs. 12 and 13. We note from these that, in the range of parameters we discuss our model, we have the case that the following holds.

$$Max[Re[W_k(x)]] = Re[W_0(x)] = Re[W_{-1}(x)],$$

and

$$Min[|Im[W_k(x)]|] = Im[W_0(x)] = -Im[W_{-1}(x)].$$

Table 1 Real part of the $W_k(\hat{b\tau})$ for $-10 \le k \le 10$ and $\tau = \{0, 2, 0.5, 0.8, 1.0, 1.5, 2.0, 5.0, 10\}$ with $a = 0.01$, $b = -2.0$

k	0.2	0.5	0.8	1.0	1.5	2.0	5.0	10.0
−10	−4.98119	−4.06499	−3.59655	−3.37504	−2.97557	−2.69648	−1.88488	−1.56667
−9	−4.86717	−3.95077	−3.48225	−3.26071	−2.86119	−2.58208	−1.77043	−1.45221
−8	−4.73849	−3.82181	−3.35317	−3.13159	−2.7320	−2.45285	−1.64115	−1.32293
−7	−4.59080	−3.67370	−3.20491	−2.98326	−2.58359	−2.30439	−1.49263	−1.17441
−6	−4.41749	−3.49977	−3.03074	−2.8090	−2.40920	−2.12994	−1.31811	−0.999913
−5	−4.20779	−3.28902	−2.81962	−2.59774	−2.19774	−1.91839	−1.10652	−0.788393
−4	−3.94219	−3.02150	−2.55143	−2.32931	−1.92901	−1.64954	−0.837777	−0.519876
−3	−3.57946	−2.65445	−2.18303	−1.9604	−1.55973	−1.2802	−0.46942	−0.15237
−2	−3.00248	−2.06355	−1.58892	−1.36578	−0.965451	−0.687324	0.114606	0.426424
−1	−0.94422	−0.319003	0.010835	0.16922	0.458766	0.664004	1.27425	1.51876
0	−0.94422	−0.319003	0.010835	0.16922	0.458766	0.664004	1.27425	1.51876
1	−3.00248	−2.06355	−1.58892	−1.36578	−0.965451	−0.687324	0.114606	0.426424
2	−3.57946	−2.65445	−2.18303	−1.9604	−1.55973	−1.2802	−0.46942	−0.15237
3	−3.94219	−3.02150	−2.55143	−2.32931	−1.92901	−1.64954	−0.837777	−0.519876
4	−4.20779	−3.28902	−2.81962	−2.59774	−2.19774	−1.91839	−1.10652	−0.788393
5	−4.41749	−3.49977	−3.03074	−2.8090	−2.40920	−2.12994	−1.31811	−0.999913
6	−4.59080	−3.67370	−3.20491	−2.98326	−2.58359	−2.30439	−1.49263	−1.17441
7	−4.73849	−3.82181	−3.35317	−3.13159	−2.7320	−2.45285	−1.64115	−1.32293
8	−4.86717	−3.94952	−3.48225	−3.26071	−2.86119	−2.58208	−1.77043	−1.45221
9	−4.98119	−3.95077	−3.59655	−3.37504	−2.97557	−2.69648	−1.88488	−1.56667
10	−5.08353	−4.16749	−3.69911	−3.47763	−3.07819	−2.79913	−1.98757	−1.66936

Table 2 Imaginary part of the $W_k(\hat{b}\tau)$ for $-10 \leq k \leq 10$ and $\tau = \{0, 2, 0.5, 0.8, 1.0, 1.5, 2.0, 5.0, 10\}$ with $a = 0.01$, $b = -2.0$

k	0.2	0.5	0.8	1.0	1.5	2.0	5.0	10.0
−10	−58.0338	−58.0496	−58.0576	−58.0614	−58.0683	−58.0731	−58.0870	−58.0925
−9	−51.7425	−51.7601	−51.7691	−51.7734	−51.7811	−51.7865	−51.8021	−51.8083
−8	−45.4492	−45.4692	−45.4795	−45.4844	−45.4931	−45.4992	−45.5171	−45.5240
−7	−39.1532	−39.1764	−39.1883	−39.1939	−39.2041	−39.2112	−39.2319	−39.2400
−6	−32.8531	−32.8807	−32.8948	−32.9016	−32.9137	−32.9221	−32.9467	−32.9564
−5	−26.5463	−26.5804	−26.5979	−26.6062	−26.6212	−26.6316	−26.6621	−26.6740
−4	−20.2279	−20.2724	−20.2953	−20.3061	−20.3257	−20.3394	−20.3793	−20.3949
−3	−13.8849	−13.9491	−13.9823	−13.9980	−14.0264	−14.0463	−14.1039	−14.1264
−2	−7.47189	−7.58847	−7.64917	−7.67794	−7.72972	−7.76570	−7.86855	−7.9079
−1	−0.406786	−1.33649	−1.57766	−1.67168	−1.81798	−1.90602	−2.11338	−2.17949
0	0.406786	1.33649	1.57766	1.67168	1.81798	1.90602	2.11338	2.17949
1	7.47189	7.58847	7.64917	7.67794	7.72972	7.76570	7.86855	7.9079
2	13.8849	13.9491	13.9823	13.9980	14.0264	14.0463	14.1039	14.1264
3	20.2279	20.2724	20.2953	20.3061	20.3257	20.3394	20.3793	20.3949
4	26.5463	26.5804	26.5979	26.6062	26.6212	26.6316	26.6621	26.6740
5	32.8531	32.8807	32.8948	32.9016	32.9137	32.9221	32.9467	32.9564
6	39.1532	39.1764	39.1883	39.1939	39.2041	39.2112	39.2319	39.2400
7	45.4492	45.4692	45.4795	45.4844	45.4931	45.4992	45.5171	45.5240
8	51.7425	51.7601	51.7691	51.7734	51.7811	51.7865	51.8021	51.8083
9	58.0338	58.0496	58.0576	58.0614	58.0683	58.0731	58.0870	58.0925
10	64.3238	64.338	64.3452	64.3487	64.3549	64.3592	64.3718	64.3767

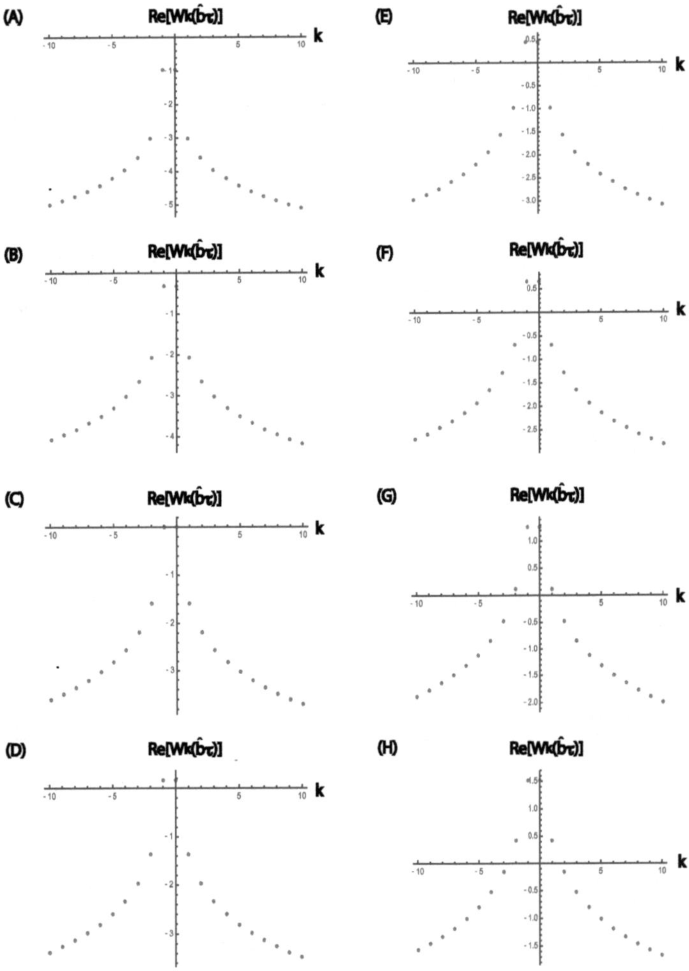

Fig. 12 Plots of Table 1. The values of the delays τ are **A** 0.2, **B** 0.5, **C** 0.8, **D** 1.0, **E** 1.5, **F** 2.0, **G** 5, **H** 10

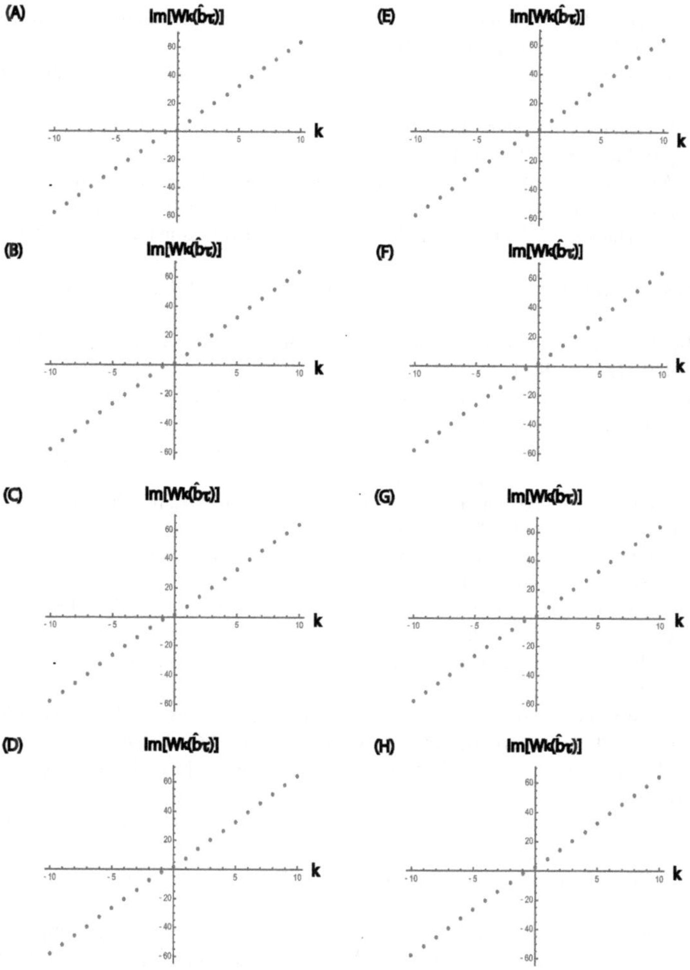

Fig. 13 Plots of Table 2. The values of the delays τ are (A)0.2, (B)0.5, (C)0.8, (D)1.0, (E)1.5, (F)2.0, (G)5, (H)10

References

an der Heiden, U.: Delays in physiological systems. J. Math. Biol. **8**, 345 (1979)

Bellman, R., Cook, K.: Differential-Difference Equations. Academic, New York (1963)

Benzi, R., Parisi, G., Sutera, A., Vulpiani, a.: Stochastic resonance in climatic change. Tellus **84**, 10–15 (1982)

Bulsara, A.R., Gammaitoni, L.: Tuning in to noise. Phys. Today **49**(3), 39–45 (1996)

Busenberg, S.N., Cooke, K.L.: Stability conditons for linear non-autonomous delay differential equations. Q. Appl. Math. **10**, 295–396 (1984)

Cabrera, J.L., Milton, J.G.: On-off intermittency in a human balancing task. Phys. Rev. Lett. **89**, 158702 (2002)

Cantisán, J., Seoane , J.M., Sanjuán, M.A.F.: Transient chaos in time-delayed systems subjected to parameter drift. J. Phys. Complex **2**, 025001 (2021)
Collins, J.J., Chow, C.C., Imhoff, T.T.: Stochastic resonance without tuning. Nature **376**, 236–238 (1995)
Corless, R.M., Gonnet, G.H., Hare, D.E.G., Jeffrey, D.J., Knuth, D.E.: On the Lambert W function. Adv. Comput. Math. **5**, 329–359 (1996)
Ford, N.J., Verduyn Lunel, S.M.: Characterising small solutions in delay differential equations through numerical approximations. Appl. Math. Comput. **131**, 253–270 (2002)
Gammaitoni, L., Hänggi, P., Jung, P., Marchesoni, F.: Stochastic resonance. Rev. Modern Phys. **70**, 223 (1998)
Glass, L., Mackey, M.C.: From Clocks to Chaos: The Rhythms of Life. Princeton University Press, Princeton, New Jersey (1988)
Glass, L., Beuter, A., Larocque, D.: Time delays, oscillations, and chaos in physiological control systems. Math. Biosci. **90**, 111 (1988)
Győri, I., Horváth, L.: Sharp estimation for the solution of delay differential and hallway type inequalities. Discrete Contin. Dyn. Syst. **37**, 3211–3242 (2017)
Hayes, N.D.: Roots of the transcendental equation associated with a certain difference-differential equation. J. Lond. Math. Soc. **25**, 226 (1950)
Insperger, T.: Act-and-wait concept for continuous-time control systems with feedback delay. IEEE Trans. Control Sys. Technol. **14**, 974 (2006)
Küchler, U., Mensch, B.: Langevin's stochastic differential equation extended by a time-delayed term. Stoch. Stoch. Rep. **40**, 23 (1992)
Küchler, U., Mensch, B.: Langevin's stochastic differential equations extended by a time-delayed term. Stochast. Stochast. Rep. **40**, 23 (1992)
L. LI. Ming, Stability for linear non autonomous delay differential equations. Math. Comput. Model. **10**, 67–74 (1990)
Longtin, A., Milton, J.G.: Insight into the transfer function, gain and oscillation onset for the pupil light reflex using delay-differential equations. Biol. Cybern. **61**, 51 (1989)
Longtin, A., Bulsara, A., Moss, F.: Time-interval sequences in bistable systems and the noise-induced transmission of information by sensory neurons. Phys. Rev. Lett. **67**, 656 (1991)
Mackey, M.C., Glass, L.: Oscillation and chaos in physiological control systems. Science **197**, 287 (1977)
Massoller, C.: Noise-induced resonance in delayed feedback systems. Phys. Rev. Lett. **88**, 034102 (2002)
Milton, J., Cabrera, J.L., Ohira, T., Tajima, S., Tonosaki, Y., Eurich, C.W., Campbell, S.A.: The time-delayed inverted pendulum: implications for human balance control. Chaos **19**, 026110 (2009)
Milton, J., Naik, P., Chan, C., Campbell, S.A.: Indecision in neural decision making models. Math. Model. Nat. Phenom. **5**, 125 (2010)
Misono, M., Todo, T., Miyakawa, K.: Coherence resonance in a schmitt-trigger inverter with delayed feedback. J. Phys. Soc. Jpn. **78**(1), 014802 (2009)
Moss, F., Wisenfeld, K.: The benefits of background noise. Sci. Am. **273**(August), 66 (1995)
Ohira, K., Ohira, T.: Delay, resonance and the Lambert W function. arxiv:2301.13448 (accepted and to appear in Springer Proceedings in Physics, 2023)
Ohira, T., Sato, Y., Cowan, J.D.: Resonance in a stochastic neuron model with delayed interaction. In: Solla, S.A., Leen, T.K., Muller, K.-R. (eds.) Advances in Neural Information Processing Systems, vo. 12, pp. 314–320. MIT Press (2000)
Ohira, T., Sato, Y.: Resonance with noise and delay. Phys. Rev. Lett. **82**, 2811 (1999)
Ohira, K.: An exact solution for a non-autonomous delay differential equations (2024). arXiv:2411.11402
Ohira, T.: Oscillatory correlation of delayed random walks. Phys. Rev. E **55**, R1255 (1997)
Ohira, T.: Delayed Stochastic Resonance with a 1-D ring. XXPhys. A **314**, 146 (2002)
Ohira, K.: Resonating delay equation. Eur. Phys. Lett. **137**, 23001 (2022)

Ohira, T., Milton, J.G.: Delayed random walks. Phys. Rev. E **52**, 3277 (1995)
Ohira, K., Ohira, T.: Delayed dynamics with transient resonating oscillations. J. Phys. Soc. Jpn. **92**, 064002 (2023)
Ohira, K., Ohira, T.: Transient reviving dynamics with an exact solution for delay differential equations. J. Phys. Soc. Jpn. **93**, 094801 (2024)
Ohira, K., Ohira, T.: Solving a delay differential equation through fourier transform. Phys. Lett. A **531**, 130138 (2025)
Ohira, T., Sato, Y.: Resonance with noise and delay. Phys. Rev. Lett. **82**, 2811–2815 (1999)
Ohira, T., Yamane, T.: Delayed stochastic systems. Phys. Rev. E **61**, 1247 (2000)
Ohira, T., Yamane, T.: Delayed stochastic systems. Phys. Rev. E **61**, 1247–1257 (2000)
Pakdaman, K., Grotta-Ragazzo, C., Malta, C.P.: Transient regime duration in continuous-time neural networks with delay. Phys. Rev. E **58**, 3623 (1998)
Priplata, A., Niemi, J., Salen, M., Harry, J., Lipsitz, L.A., Collins, J.J.: Noise-enhanced human balance control. Phys. Rev. Lett. **89**, 23 (2002)
Pusenjak, R.: Application of Lambert function in the control of production systems with delay. Int. J. Eng. Sci. **6**, 28–38 (2017)
Rose, J.E., Brugge, J.F., Anderson, D.J., Hind, J.E.: Phase-locked response to low-frequency tones in single auditory nerve fibers of the squirrel monkey. J. Neurophysiol. **30**, 769–793 (1967)
Sakurai, J.J.: Modern Quantum Mechanics. Benjamin/Cummings, Menlo Park, California (1985)
Shinozaki, H., Mori, T.: Robust stability analysis of linear time delay system by Lambert W function. Automatica **42**, 1791–1799 (2006)
Smith, H.: An Introduction to Delay Differential Equations with Applications to the Life Sciences. Springer, New York (2010)
Stepan, G.: Retarded Dynamical Systems: Stability and Characteristic Functions. Wiley, New York (1989)
Stepan, G., Insperger, T.: Stability of time-periodic and delayed systems: a route to act-and-wait control. Ann. Rev. Control **30**, 159 (2006)
Szydlowski, M., Krawiec, A.: The Kaldor-Kalecki model of business cycle as a two-dimensional dynamical system. J. Nonlinear Math. Phys. **8**, 266 (2010)
Taylor, S.R., Campbell, S.A.: Approximating chaotic saddles for delay differential equations. Phys. Rev. E **75**, 046215 (2007)
Tsimring, L.S., Pikovsky, A.: Noise-induced dynamics in bistable systems with delay. Phys. Rev. Lett. **87**, 250602 (2001)
Wiesenfeld, K., Moss, F.: Stochastic resonance and the benefits of noise: from ice ages to crayfish and SQUIDs. Nature **373**, 33–36 (1995)

Periodic Solutions of a Delay Differential Equation with a Periodic Multiplier

Anatoli Ivanov, Bernhard Lani-Wayda, and Sergiy Shelyag

Abstract A simple non-autonomous scalar differential equation with delay, exponential decay, nonlinear negative feedback and a periodic multiplicative coefficient is considered. It is shown that stable slowly oscillating periodic solutions with the period of the feedback coefficient, and also with the double period of the feedback coefficient exist. The periodic solutions are built explicitly in the case of piecewise constant feedback function and the periodic coefficient. The periodic dynamics are shown to persist under small perturbations of the equation which make it smooth. The results are confirmed and illustrated by numerical simulations.

Keywords Non-autonomous scalar delay differential equation · Exponential decay · Nonlinear negative feedback · Periodic multiplicative coefficient · Slowly oscillating periodic solutions · Stable periodic dynamics

1 Introduction

The simple-looking scalar delay differential equation

$$x'(t) = -\mu x(t) + f(x(t-\tau)), \quad f \in C^0(\mathbb{R}, \mathbb{R}), \ \mu \geq 0, \ \tau > 0 \tag{1}$$

is a paradigm for the rich and complex dynamics that such equations exhibit, as well as for a wide range of applications in various fields of science, engineering,

---

A. Ivanov (✉)
Pennsylvania State University, Dallas, USA
e-mail: aivanov@psu.edu

B. Lani-Wayda
Justus-Liebig-Universität, Giessen, Germany
e-mail: Bernhard.Lani-Wayda@math.uni-giessen.de

S. Shelyag
Flinders University, Adelaide, Australia
e-mail: Sergiy.Shelyag@flinders.edu.au

E. Braverman and A. Swishchuk (eds.), *Delay and Stochastic Differential Equations*, ICIAM2023 Springer Series 12, https://doi.org/10.1007/978-981-95-7310-3_2

and other real world phenomena. Unlike the corresponding scalar autonomous ordinary differential equation without delay, it can exhibit various dynamical behaviors such as sustained and periodic oscillations about equilibria, and complex (chaotic) dynamics. Some examples of these dynamics and further references can be found in e.g. An der Heiden and Walther (1983), An der Heiden and Mackey (1982), Glass and Mackey (1988), Ivanov and Sharkovsky (1992), Lani-Wayda (1995a), Mackey and Glass (1977), Peters (1983), Walther (1981b). The broad spectrum of dynamical patterns and the intrinsic presence of delay in many applications make this equation a simple but adequate model to describe a range of various processes in biological sciences. Among those are the well-known blood production and functioning models of Mackey-Glass (1977) and Lasota-Wazewska (1976), population models of Nicholson's type (Gurney et al. 1980; Perez et al. 1978), other physiological and population dynamics models (Glass and Mackey 1988; Kuang 1993). The broad range of applications of this equation is described in multiple sources; here we mention a few books which focus on the applications as well as on the theoretical basics of such equations (Erneux 2009; Kuang 1993; Smith 2011). The fundamentals of the theory of this and more general delay differential equations can be found in e.g. monographs Diekmann et al. (1995), Hale (1977), Hale and Verduyn Lunel (1993).

Extended models take into account external influences—in particular, periodic factors that can be naturally present in some components of the system, such as e.g. circadian rhythms, periodic medication intake, or seasonal fluctuations in biological models. Such considerations lead to non-autonomous delay differential equations with periodic time dependence, and a special case is the following generalization of DDE (1):

$$x'(t) = -\mu(t)x(t) + a(t)f(x(t - \tau(t))), \quad f \in C^0(\mathbb{R}, \mathbb{R}), \tag{2}$$

where all functions $\mu(t)$, $a(t)$, $\tau(t)$ are continuous and T-periodic. In this paper we focus on a special case of DDE (2) where the time delay τ and the decay rate μ are constants, while the delayed feedback input is periodic through the multiplicative coefficient $a(t)$ with the nonlinearity $f(x)$.

We note that there is a good number of publications on scalar delay differential equations with periodic dependence on time in coefficients and delays. However, in most of them the time dependence is such that the corresponding equation does not possess a constant equilibrium any more; see e.g. Ai (2024), Amster and Idels (2013), Faria (2017), Friedman and Wu (1992) and further references therein (our Eq. (2) always has the constant solution $x(t) \equiv 0$). Some of those periodic equations do admit the constant (zero) solution; however, in those papers no distinction can be made between nontrivial periodic solutions and the existing zero solution; see e.g. Li and Kuang (2001). Periodic solutions are frequently obtained from topological methods such as Schauder's fixed point theorem (see e.g. Faria 2017) or the Gaines-Mawhin coincidence degree (see e.g. Li and Kuang 2001), they do not yield results on stability of the periodic solutions. In previous papers, the periodic influence often appears in 'additive' form like $\ldots + g(t, x(t), x(t-r))$ (e.g., Liu et al. 2005), although multiplication with periodic coefficients has also been studied, e.g. in Faria (2017) or Jiang et al. (2002). Recently, a partial case of Eq. (2), when $\mu \equiv 0$, $\tau(t) = \tau_0 =$ constant, was considered in Ivanov and Shelyag (2022, 2024, 2025).

In the present paper we focus on a simple special case of DDE (2) with the normalized delay $\tau = 1$:

$$x'(t) = -\mu x(t) + a(t) f(x(t-1)), \tag{3}$$

Here we can provide detailed information on the shape of solutions, criteria for existence of periodic solutions, and their stability.

In Sect. 2 we explicitly compute solutions for the case of piecewise constant f and a, and derive conditions for the existence and stability/instability of periodic solutions with the same period as the coefficient. In Sect. 3, a similar approach yields solutions with twice the coefficient period as their minimal period.

Focusing on the single-period case, a smoothing procedure is described in Sect. 4. This gives nearby equations within a C^1-framework, preserving periodic solutions and their stability or instability. Such behavior is then robust under arbitrary C^1-small perturbations.

We finally note that the global dynamics of both the discontinuous and the smoothed versions of Eq. (3) may also contain rapidly oscillating periodic solutions, connecting orbits, other dynamics, as it is already the case for the autonomous equation $\dot{x}(t) = -\mu \cdot \text{sign}(x(t-1))$. Comprehensive study of the complete solution behaviors is beyond the scope of this paper (we do not talk about an attractor here, since the equation is non-autonomous, but one could consider e.g. the attractor of the period map). We mention in this context the challenging problem of proving that the domain of attraction into 'slowly oscillating behavior' is dense in the state space $C^0([-1, 0], \mathbb{R})$, which to our knowledge is still an open problem for autonomous equations with negative feedback, and is only proved for special cases (see e.g. Corollary 4, p. 137 in Walther 1981a).

We remark that in the case of stable periodic solutions, the stability results that we obtain are 'more than local'; e.g., in the following sense: the set D^+ (see Sect. 3.4) is a 'large' attracting neighborhood of the initial segment of the periodic solution.

2 Explicit Periodic Solutions

In this section we build explicit forward (for $t \geq 0$) solutions to Eq. (3), where the piecewise constant functions $f = f_0(x)$ and $a = A_0(t)$ are defined as follows:

$$f_0(x) = -\text{sign}(x) = \begin{cases} -1, & \text{if } x > 0 \\ 0, & \text{if } x = 0 \\ +1, & \text{if } x < 0, \end{cases}$$

and

$$A_0(t) = \begin{cases} a_1 > 0, & \text{if } 0 \leq t < p_1 \\ a_2 > 0, & \text{if } p_1 \leq t < p_1 + p_2 := T \\ \text{periodic extension outside } [0, T) \text{ for all } t \in \mathbb{R}. \end{cases} \tag{4}$$

.

In this section the initial functions which are of definite sign on the initial time interval $[-1, 0]$ will be most important.

Note that the solution to the initial value problem $\dot{x}(t) = -\mu x(t) + A, x(t_0) = x_0$, A-const, is given by $x(t) = A/\mu + (x_0 - A/\mu)\exp\{-\mu(t - t_0)\}$.

2.1 Periodic Solutions with the Same Period as the Coefficient a

In this subsection we are dealing with slowly oscillating solutions (i.e., the distance between consecutive zeroes is larger than the delay 1) of the form as shown in Fig. 1. That is, for given $h > 0$, the first two zeros t_1 and t_2 of the solution satisfy $t_1 + 1 < t_2 < p_1$ and $x(t) < 0, \forall t \in (t_1, t_2]$. In other words, the first negative semi-cycle of the solution $x = x(t, h)$ happens on the first constant segment $[0, p_1]$ of the coefficient $A(t)$. The solution remains positive on the interval $(t_2, T]$, $T = p_1 + p_2$.

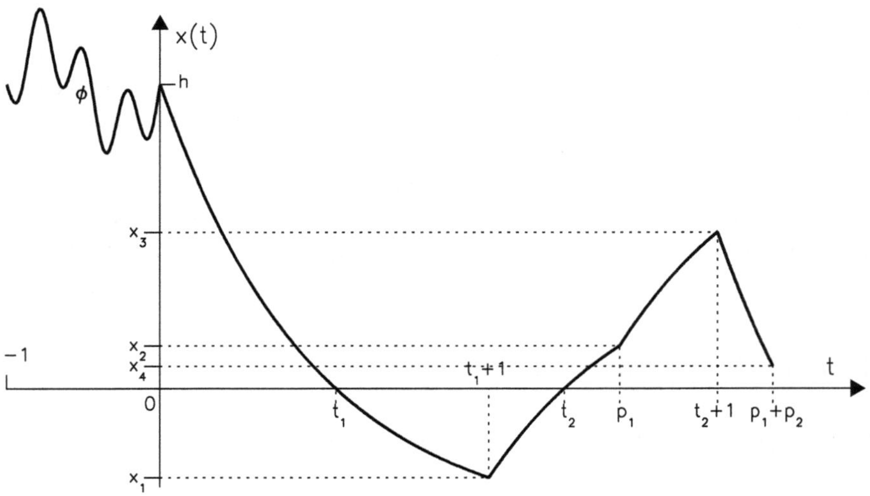

Fig. 1 Slowly oscillating solution of Eq. (3) with $h > 0$ and $x(p_1 + p_2) > 0$

Using the piecewise constant form of f_o and A_o with appropriate values on the interval $[0, p_1]$, one easily calculates the following results

$$t_1 = \frac{1}{\mu} \ln \frac{h\mu + a_1}{a_1}, \quad x_1 = x(t_1 + 1) = -\frac{a_1}{\mu} + \frac{a_1}{\mu} \exp\{-\mu\} < 0.$$

The second zero t_2 is calculated as

$$t_2 = t_1 + 1 + \frac{1}{\mu} \ln \frac{a_1 - \mu x_1}{a_1}.$$

The value x_2 is found from the condition $x_2 := x(p_1)$ as following

$$x_2 = x(p_1) = \frac{a_1}{\mu} + \left(x_1 - \frac{a_1}{\mu}\right) \exp\{-\mu[p_1 - (t_1 + 1)]\} > 0.$$

On the interval $[p_1, t_2 + 1]$ the value of x_3 at the endpoint is found as

$$x_3 = x(t_2 + 1) = \frac{a_2}{\mu} + \left(x_2 - \frac{a_2}{\mu}\right) \exp\{-\mu(t_2 + 1 - p_1)\}.$$

Finally, the value of x_4 is defined as $x_4 := x(T) = x(p_1 + p_2)$ and is found from integration on $[t_1 + 1, p_1 + p_2]$ as

$$x_4 = x(p_1 + p_2) = -\frac{a_2}{\mu} + \left(x_3 + \frac{a_2}{\mu}\right) \exp\{-\mu[p_1 + p_2 - (t_2 + 1)]\}.$$

In order for this solution to be slowly oscillating and of the desired shape as in Fig. 1 we impose the following restrictions on the zeros t_1, t_2 and parametric values a_1, a_2, p_1, p_2: $0 < t_1 < t_1 + 1 < t_2 < p_1$ and $p_1 - t_2 < 1$ (so that $p_1 < t_2 + 1 < p_1 + p_2$) and $x_4 > 0$. These conditions are numerically verified to be valid for parameter values as given in Table 1.

It is easily seen that the value of $x_4 > 0$ does not depend on the particular values of the initial function $\phi(s)$ on the interval $[-1, 0)$; it is entirely determined by the positive value $\phi(0) := h > 0$ only.

Introduce the one-dimensional map F by $F : h \mapsto x_4$. It is immediate that the fixed points h_* of this map give rise to periodic solutions of Eq. (3) with period $T = p_1 + p_2$. Namely, the initial state φ of such a periodic solution is obtained by starting any solution x with $x(0) = h_*$ and $x > 0$ on $[-1, 0]$, and then taking $\varphi := x_T = x(p_1 + p_2 + \cdot)|_{[-1,0]}$. By using the above formulas for x_1, x_2, x_3, x_4 and t_1, t_2 and by an inductive calculation one arrives at the following expression for the map F:

$$x_4 = F(h) = m \cdot h + b, \tag{5}$$

Table 1 Samples of parametric values for which Eq. (3) has a stable T-periodic solution

a_1	a_2	p_1	p_2	$\|m\|$	b	μ	T
2	0.1	3	1	0.72	1.76	0.1	4
2	0.5	3	2.5	0.31	0.43	0.1	5.5
3	1	3	1	0.21	1.69	0.1	4
3	0.1	3	7	0.41	1.03	0.1	10
1	0.5	3	0.5	0.26	0.3	0.1	3.5
4	1	3	2.5	0.31	0.87	0.1	5.5
6	0.5	3	5	0.44	1.74	0.1	8
6	2	3	1	0.21	3.38	0.1	4
1	0.17	3	0.5	0.18	0.61	0.5	3.5
1	0.25	3	1	0.06	0.3	0.5	4
2	0.1	3	0.5	0.33	1.43	0.5	3.5
2	0.1	3	2.5	0.12	0.4	0.5	5.5
2	0.1	3	3	0.1	0.27	0.1	6
2	0.5	3	1	0.06	0.6	0.5	4
3	1	2	1	0.05	0.73	0.5	3
2	1	2	1	0.38	0.13	1	3
3	0.25	2	1	0.12	0.40	1	3
4	0.1	2	2.5	0.04	0.05	1	4.5
4	0.25	2	1	0.15	0.55	1	3
4	3	2	1	0.69	0.09	1	3
6	0.1	2	1	0.2	0.87	1	3
6	0.1	2	3	0.03	0.03	1	5

where

$$m = \left[2\frac{a_2}{a_1} - \exp\{-\mu\}\right] \cdot (2 - \exp\{-\mu\}) \exp\{-\mu(p_1 + p_2 - 2)\}, \qquad (6)$$

and

$$\begin{aligned} b =& \frac{2a_2 - a_1 \exp\{-\mu\}}{\mu} (2 - \exp\{-\mu\}) \exp\{-\mu(p_1 + p_2 - 2)\} \\ &+ \frac{a_1 - a_2}{\mu} \exp\{-\mu p_2\} - \frac{a_2}{\mu} \end{aligned} \qquad (7)$$

Note that the outcome for map F being dependent only on the value $h = x(0)$ is a direct consequence of the fact that $f_0 = -$ sign is constant on $(-\infty, 0)$ and on $(0, \infty)$. This kind of 'loss of information' is typical for such piecewise constant models; compare e.g. Refs. An der Heiden and Walther (1983), Walther (1981b), and

Hale and Lin (1986), where rapidly oscillating solutions of autonomous equations are considered, and the single relevant parameter is the distance between consecutive zeroes.

Proposition 2.1 *Suppose that the parameters a_1, a_2, p_1, p_2 are such that the inequalities $|m| < 1$ and $b > 0$ are satisfied for map (5). Then there exists a unique solution $h_* > 0$ to the equation $F(h) = h$ which is given by the expression $h_* = b/(1-m)$, where m and b are defined by (6) and (7), respectively.*

Corollary 2.2 *Under the assumptions of Proposition 2.1 DDE (3) has a stable slowly oscillating periodic solution $x : \mathbb{R} \to \mathbb{R}$ of period T. Its initial state $\varphi = x_0$ is obtained, as described above, e.g. by starting with the constant initial function $\psi \equiv h_*$ and applying the period map. The function $-x$ is then also a periodic solution, as follows from oddness of the right hand side of the equation.*

2.2 *Periodic Solutions with Period Twice the Period of the Coefficient a*

In this subsection we are dealing with slowly oscillating solutions of the form as shown in Fig. 2. That is, for given $h > 0$, the first zero t_1 of the solution satisfies $t_1 < p_1$, $p_1 < t_1 + 1 < T$, and $x(t) < 0$, $\forall t \in (t_1, T]$ with $T - p_1 > 1$. This will be guaranteed if one assumes that $x_2 < 0$ and $x_3 < 0$. Using the piecewise constant form of f_o and A_o with appropriate values on the interval $[0, p_1]$, one easily derives the following calculations

$$t_1 = \frac{1}{\mu}\ln\frac{h\mu + a_1}{a_1}, \quad x_1 = x(p_1) = -\frac{a_1}{\mu} + \left(h + \frac{a_1}{\mu}\right)\exp\{-\mu p_1\}.$$

On the interval $[p_1, t_1 + 1]$ one finds:

$$x_2 = x(t_1 + 1) = -\frac{a_2}{\mu} + \left(x_1 + \frac{a_2}{\mu}\right)\exp\{-\mu(t_1 + 1 - p_1)\} < 0.$$

On the interval $[t_1 + 1, p_1 + p_2]$ the value of x_3 is found as

$$x_3 = x(p_1 + p_2) = \frac{a_2}{\mu} + \left(x_2 - \frac{a_2}{\mu}\right)\exp\{-\mu[p_1 + p_2 - (t_1 + 1)]\}.$$

The value of $x_3 < 0$ does not depend on the particular values of the initial function $\phi(s)$ on the interval $[-1, 0)$; it is entirely determined by the positive value $\phi(0) = h > 0$ only.

Introduce the one-dimensional map Φ_1 by $\Phi_1 : h \mapsto h_1 := x_3$.

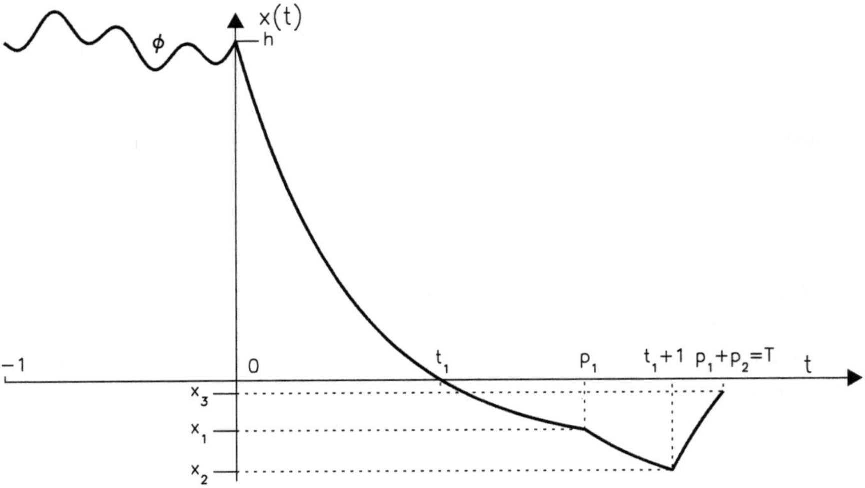

Fig. 2 Slowly oscillating solution of Eq. (3) with $h > 0$ and $x(p_1 + p_2) < 0$

Likewise, due to symmetry considerations, for every initial function ψ satisfying $\psi(s) < 0\ \forall s \in [-1, 0]$ with $\psi(0) := h_1$ the corresponding solution $x = x(t, \psi), t \geq 0$, can be constructed with $x(p_1 + p_2) := h_2 > 0$. This defines an analogous one-dimensional map $\Phi_2 : h_1 \mapsto h_2$—in fact, $\Phi_2(h) = -(\Phi_1(-h))$

Fixed points h_* of the composite one-dimensional map $\Phi = \Phi_2 \circ \Phi_1$ give rise to slowly oscillating periodic solutions of Eq. (3). In particular, the positive solutions h_* of the equation $\Phi_1(h) = -h, h > 0$, provide such fixed points of the map Φ (in general, there may exist additional fixed points of map Φ which are not solutions of this equation). We shall derive next some of the conditions on the parameters a_1, a_2, p_1, p_2 which guarantee the existence of the positive solutions to equation $\Phi_1(h) = -h$.

By using the recursive equations above for the values of x_1, x_2 and x_3 one finds that

$$x_3 = x(T) = x(p_1 + p_2) = \Phi_1(h) = k \cdot h + d, \tag{8}$$

where

$$k = \left(1 - 2\frac{a_2}{a_1}\exp\{\mu\}\right)\exp\{-\mu(p_1 + p_2)\} \tag{9}$$

and

$$d = \frac{a_2 - a_1}{\mu} \exp\{-\mu p_2\} + \frac{a_1}{\mu} \exp\{-\mu(p_1 + p_2)\} + \frac{a_2}{\mu} \left[1 - 2\exp\{-\mu(p_1 + p_2 - 1)\}\right]. \tag{10}$$

Proposition 2.3 *Suppose that the parameters a_1, a_2, p_1, p_2 are such that the inequalities $|k| < 1$ and $d < 0$ are satisfied. Then there exists a unique solution $h_* > 0$ to the equation $\Phi_1(h) + h = 0$ which is given by the expression $h_* = -d/(k+1)$, where k and d are defined by (9) and (10), respectively (Table 2).*

Corollary 2.4 *Under the assumptions of Proposition 2.3 DDE (3) has a symmetric slowly periodic solution of period $2T$ satisfying $x(t+T) \equiv -x(t), \forall t \in \mathbb{R}$. As above, the solution can be obtained by starting with the initial function $\varphi \equiv h_*$.*

Table 2 Samples of parameters for which Eq. (3) has stable periodic solutions with double period $2T$

a_1	a_2	p_1	p_2	$\|k\|$	d	μ	$2T$
1	0.17	1	4	0.38	−0.09	0.1	10
2	0.5	1	2.5	0.32	−0.38	0.1	7
2	0.25	1	5	0.40	−0.17	0.1	12
4	2	0.5	1	0.09	−1.72	0.1	3
4	0.1	1	7	0.43	−1.39	0.1	16
6	0.25	0.5	5	0.52	−0.95	0.1	11
2	0.5	0.5	1	0.08	−0.49	0.5	3
2	0.25	1	2.5	0.10	−0.09	0.5	7
2	0.17	1	3	0.1	−0.09	0.5	8
3	0.25	0.5	2.5	0.16	−0.11	0.5	6
4	0.1	1	5	0.05	−0.08	0.5	12
4	1	0.5	1	0.08	−0.98	0.5	3
2	0.5	0.5	1	0.08	−0.21	1	3
2	0.1	1	2.5	0.02	−0.01	1	7
3	0.17	0.5	1	0.16	−0.41	1	3
4	1	0.5	1	0.08	−0.42	1	3
4	0.1	1	3	0.02	−0.03	1	8
6	1	0.5	1	0.02	−0.71	1	3

3 Smoothing Nonlinearities

We demonstrate how to get a smooth dynamical system close to the one generated by Eq. (3) with piece-wise constant functions in Sect. 2 by smoothing the nonlinearities involved. In a small δ-neighborhood of the discontinuity set of functions $f = f_0(x) = -\operatorname{sign}(x)$ and $a = A_0(t)$ given by (4) they are replaced by continuous segments (thus making the nonlinearities continuous) while retaining their piece-wise constant values everywhere else outside the δ-neighborhood. Such a smoothing procedure for partial cases of Eq. (3) was applied in papers Ivanov and Shelyag (2024), Ivanov and Shelyag (2022) where the connecting lines are straight line segments (making functions f and a continuous only). As it is shown in this section the smoothing procedure in the small δ-neighborhood can be such that the resulting continuous functions $\tilde{f}$ and $\tilde{a}$ are C^1-smooth (or even of class C^∞). Since $\tilde{f}$ and $\tilde{a}$ differ from f and a respectively only on small δ-neighborhood of the discontinuity set for the latter, the natural expectation is that the dynamical properties of the smoothed Eq. (3) are close to those with the piece-wise continuous functions. This fact is rigorously established in the current section (see Theorems 3.5 and 3.6 below). The main purpose of changing the discontinuous functions to C^1-functions is to get into the framework of 'standard' C^1 dynamical systems theory. Then, as it is implemented for example in papers Walther (1981b) and Hale and Lin (1986), the principles of stability and instability via linearization, or invariant manifold theory, become applicable.

For given functions $\tilde{a}, \tilde{f} : \mathbb{R} \to \mathbb{R}$ and $\varphi \in C^0([-1, 0], \mathbb{R})$, we denote by $x^\varphi(\cdot, \tilde{a}, \tilde{f})$ the solution to the equation $x'(t) = -\mu x(t) + \tilde{a}(t)\tilde{f}(x(t-1))$ with initial state $x_0 = x_{|[-1,0]} = \varphi$. As in previous sections, 'solution' is to be understood in the sense that $x = x^\varphi(\cdot, \tilde{a}, \tilde{g})$ is continuous on an interval $[-1, t^+)$ and differentiable at all times $t \in [0, t^+)$ at which a is continuous and where $x(t-1) \neq 0$, and satisfies the equation for these t. (At $t = 0$, this refers to right-side derivative). Since for our equations the forward solution depends only on $h = \varphi(0)$, we frequently also write $x^h(\dots.)$ instead of $x^\varphi(\dots.)$. Obviously such a solution satisfies the integral equation:

$$x(t) = e^{-\mu(t-t_0)} \cdot x(t_0) + \int_{t_0}^{t} e^{-\mu(t-s)}\tilde{a}(s)\tilde{f}(x(s-1))\, ds \quad (t \in [t_0, t^+)).$$

3.1 Preparations

We want to replace the step functions a given by (4) and $f = -\operatorname{sign}$ by nearby functions which are at least of class C^1, preserving as much as possible of the original dynamics. We need some technical preparation for this:

Proposition 3.1 *Assume that $\alpha, \beta \in \mathbb{R}$ with $\alpha \cdot \beta < 0$, that a function x satisfies $x(t) = \alpha + \beta e^{-\mu t}$ on an open interval I, and $x(t_0) = 0$ for some $t_0 \in I$. Let $\delta > 0$ be so small that $[-\delta, \delta] \subset x(I)$.*
(a) Then, with $\theta_+(\delta) := \frac{-1}{\mu} \ln(1 - \frac{\delta}{\alpha})$, $\theta_-(\delta) := -\frac{1}{\mu} \ln(1 + \frac{\delta}{\alpha})$, one has

$$x(t_0 + \theta_-(\delta)) = -\delta, \ x(t_0 + \theta_+(\delta)) = \delta,$$

and $\theta_\pm(\delta) \to 0$ as $\delta \to 0$, locally uniformly w.r. to α, β
(b) For a piecewise continuous function $\tilde{f} : \mathbb{R} \to \mathbb{R}$ one has

$$\int_{t_0+\theta_-(\delta)+1}^{t_0+\theta_+(\delta)+1} e^{-\mu[t_0+\theta_+(\delta)+1-s]} \tilde{f}(x(s-1))\, ds = J(\alpha, \tilde{f}), \text{ with}$$

$$J(\alpha, g) := \frac{\alpha - \delta}{\mu} \int_{-\delta}^{\delta} \frac{g(u)}{(\alpha - u)^2} du \quad \text{for } g \in \mathcal{L}^1([-\delta, \delta), \mathbb{R}).$$

(In particular, for fixed δ and μ, the value of this integral depends only on the limit value α of $\alpha + \beta e^{-\mu t}$ for $t \to \infty$, but not on β.)

Proof (For the proof, we omit the δ−dependence of $\theta_\pm$ in the notation.)

(Ad a): θ_+ is characterized by the equation $\beta \cdot [e^{-\mu(t_0+\theta_+)} - e^{-\mu t_0}] = \delta$, and $\beta e^{-\mu t_0} = -\alpha$ gives: $-\alpha[e^{-\mu\theta_+} - 1] = \delta$, or

$$\theta_+ = \frac{-1}{\mu} \ln(1 - \frac{\delta}{\alpha}).$$

The second formula is obtained analogously, and the convergence result is obvious from these formulas.

(Ad b): Note that $e^{\mu s} = \beta/(x(s) - \alpha)$ $(s \in I)$, in particular,

$$e^{-\mu(t_0+\theta_+)} = (x(t_0 + \theta_+) - \alpha)/\beta = (\delta - \alpha)/\beta.$$

Note also that $x'(s) = -\mu(x(s) - \alpha)$ $(s \in I)$. Using these identities and the substitution rule, one sees that the integral in question equals

$$\begin{aligned}
&\frac{\delta - \alpha}{\beta} \int_{t_0+\theta_-}^{t_0+\theta_+} e^{\mu s} \tilde{f}(x(s))\, ds \\
&= \frac{(\delta - \alpha)}{\beta} \int_{t_0+\theta_-}^{t_0+\theta_+} \frac{\beta}{x(s) - \alpha} \tilde{f}(x(s)) \cdot \frac{x'(s)}{-\mu(x(s) - \alpha)} ds \\
&= \frac{(\delta - \alpha)}{\beta} \int_{-\delta}^{\delta} \frac{\beta}{-\mu(u - \alpha)^2} \tilde{f}(u)\, du = \frac{\alpha - \delta}{\mu} \int_{-\delta}^{\delta} \frac{\tilde{f}(u)}{(\alpha - u)^2} du = J(\alpha, \tilde{f}).
\end{aligned}$$

□

The main idea of our smoothing procedure is not at all new, but present in previous publications; see An der Heiden and Walther (1983), An der Heiden and Mackey (1982), Peters (1983), Walther (1981b), and Hale and Lin (1986) for papers using step functions and (except for An der Heiden and Mackey 1982; Peters 1983) smoothed versions of these as nonlinearities of delay equations. The most important point in these considerations is mentioned in Hale and Lin (1986) on the top of p. 696: The values of the solution x at times t_0 where its derivative has jump discontinuities (which arise as a consequence of the discontinuities of f or a in our situation) have to be a minimum distance away from the discontinuity points of the step function f (in our case, only zero), so that f and also the smooth version $\tilde{f}$ of f are constant in a neighborhood of these values. Then the (constant) values of $f(x(s))$ and of $\tilde{f}(\tilde{x}(s))$, where the tilde indicates smoothed nonlinearity and corresponding solution, coincide for s in a neighborhood of the time t_0 where x' has a jump discontinuity.

3.2 Smoothing $f = -$ sign to $\tilde{f}$

Consider solutions $x = x^h$ starting with $x(0) = h$ as described in Sect. 2.1, with zeroes at t_1 and t_2 (depending on h), for h-values close to the value h_* corresponding to the periodic solution as constructed in Sect. 2.1. For small enough $\delta > 0$, there exist the numbers $\theta_\pm = \theta_\pm(\delta)$ associated to the first crossing of zero at t_1 as in part a) of Proposition 3.1, and analogously $\vartheta_\pm$ associated to the second zero at t_2. (Note that then $\theta_- > 0 > \theta_+$ and $\vartheta_- < 0 < \vartheta_+$.)

Proposition 3.2 *For $\delta > 0$ sufficiently small, there exists a C^∞ function $\tilde{f} : \mathbb{R} \to \mathbb{R}$ with*

$$\tilde{f}(x) = f(x) = -sign(x) \text{ if } |x| \geq \delta, \text{ and } -1 \leq \tilde{f}(x) \leq 1 \text{ if } |x| \leq \delta, \tag{11}$$

and such that the solutions $\tilde{x} = x^\varphi(\cdot, a, \tilde{f})$ and $x = x^\varphi(\cdot, a, \tilde{f})$ (both still with the discontinuous coefficient a) with $\varphi(0) = h$, where h is close to h_*, have the following properties:

(a) Let $\theta_\pm$ and $\vartheta_\pm$ be as explained above. Then (with the h-dependent times t_1 and t_2),

$$\tilde{x} = x \quad \text{on } [-1, t_2 + 1 + \vartheta_-] \setminus [t_1 + 1 + \theta_+,\ t_1 + 1 + \theta_-].$$

(b) There exists a constant $C > 0$ such that on the 'exceptional' interval $[t_1 + 1 + \theta_+,\ t_1 + 1 + \theta_-]$ one has $|\tilde{x}(t) - x(t)| \leq C\delta$, and also

(c) $|\tilde{x} - x| \leq C\delta$ on $[t_2 + 1 + \vartheta_-, t_2 + 1 + \vartheta_+]$.

(d) On $[t_2 + 1 + \vartheta_+, p_1 + p_2]$ one has for sufficiently small values of δ and of $|h - h_*|$:

$$\tilde{x}(t) - x(t) = e^{-\mu[t-(t_2+\vartheta_+ +1)]} R(\delta),$$

where $R(\delta) := a_2 \cdot J(a_1/\mu, \tilde{f} - f)$ *is independent of* h*, and* $|R(\delta)| \leq C\delta$.

Proof Clearly there exist C^∞ functions $\tilde{f}$ with the properties in (11). For such a function and small enough $\delta > 0$ (with $\delta < x_1$, see Sect. 2.1), one has $x = \tilde{x}$ on $[-1, t_1 + \theta_+ + 1]$. We want to achieve $\tilde{x}(t_1 + \theta_- + 1) = x(t_1 + \theta_- + 1)$. In view of (3), applied with $t_0 = t_1 + \theta_+ + 1$, and the fact that $a = a_1$ is constant around $t_1 + 1$, this condition is equivalent to

$$\int_{t_1+\theta_++1}^{t_1+\theta_-+1} e^{-\mu[t_1+\theta_-+1-s]}\tilde{f}(\tilde{x}(s-1))\,ds = \int_{t_1+\theta_++1}^{t_1+\theta_-+1} e^{-\mu[t_1+\theta_-+1-s]}f(x(s-1))\,ds. \tag{12}$$

Now on the interval $[t_1 + \theta_+, t_1 + \theta_-]$ (where t_1 depends on h), solutions $x = x^h$ have the form $t \mapsto \alpha + \beta e^{-\mu t}$, where $\alpha = -a_1/\mu$,

$$\beta = (a_1/\mu)e^{\mu t_1} = \frac{a_1}{\mu} \cdot \frac{\mu h + a_1}{a_1} = h + a_1/\mu,$$

and only depend on h, see Sect. 2.1. The formulas for $\theta_\pm$ from part a) of Proposition 3.1 thus show that $\theta_\pm$ are independent of h. Further, part b) of the same Proposition shows that condition (12) is equivalent to

$$J(\alpha, \tilde{f}) = J(\alpha, -\operatorname{sign}). \tag{13}$$

The number $c = J(\alpha, -\operatorname{sign})$ on the right hand side of Eq. (13) is clearly between the numbers $J(\alpha, -\mathbf{1})$ and $J(\alpha, +\mathbf{1})$, where $\pm\mathbf{1}$ denotes the constant functions. By choosing $\tilde{f}_-$, $\tilde{f}_+$ according to (11) close enough to $-\mathbf{1}$ and $+\mathbf{1}$ in $\mathcal{L}^1([-\delta, \delta])$, respectively, one can achieve that c is also between $J(\alpha, \tilde{f}_-)$ and $J(\alpha, \tilde{f}_+)$. It follows then(using continuity of $J(\alpha, \cdot)$ with respect to the $\mathcal{L}^1$-norm) that there exists $s \in [0, 1]$ such that the function $\tilde{f} = (1-s)\tilde{f}_- + s\tilde{f}_+$ satisfies condition (13).

With such an $\tilde{f}$, one has on the interval $[t_1 + 1 + \theta_+,\ t_1 + 1 + \theta_-]$ that

$$\begin{aligned} |\tilde{x}(t) - x(t)| &= \Big|\int_{t_1+\theta_++1}^{t} \underbrace{e^{-\mu[t-s]}}_{\leq 1} a_1 \cdot \underbrace{[\tilde{f}(x(s-1)) - f(x(s-1))]}_{|\dots|\leq 2}\,ds\Big| \\ &\leq (\theta_- - \theta_+) \cdot 2|a_1| \leq C_1\delta \end{aligned} \tag{14}$$

with appropriate C_1, where the last estimate follows from the formulas for $\theta_\pm$ from Proposition 3.1. If δ is so small that $x_1 + C_1\delta < -\delta$ then for all solutions $\tilde{x}$ in question $\tilde{f}(\tilde{x}(s)) = +1 = f(x(s))$ $(s \in [t_1 + 1 + \theta_+,\ t_1 + 1 + \theta_-])$, which together with $\tilde{x}(t_1 + 1 + \theta_-) = x(t_1 + 1 + \theta_-)$ gives that also $x = \tilde{x}$ on $[t_1 + 1 + \theta_-, t_1 + 2 + \theta_-]$. This interval may or may not contain the second zero t_2 of x, but in any case we have $a_1\tilde{f}(x(s)) = a_1 f(x(s)) = +a_1$ for $s \in [t_1 + 1 + \theta_-, t_2 + \vartheta_-]$. Thus we obtain

$$x = \tilde{x} \text{ on } [t_1 + 1 + \theta_-, t_2 + 1 + \vartheta_-],$$

and (a) is proved.

Clearly, since $t_2 < p_1 < t_2 + 1$ and $\vartheta_\pm \to 0$ as $\delta \to 0$, we can assume that $t_2 + \vartheta_+ < p_1 < t_2 + 1 - \vartheta_-$. On the interval $[t_1 + 1 + \theta_-, p_1]$, which contains t_2 in its interior, and on which $x = \tilde{x}$, we then have $x(s) = \alpha_2 + \beta_2 e^{-\mu s} = \tilde{x}(s)$, with $\alpha_2 = a_1/\mu$ again independent of h and (see Sect. 2.1, and observe $x(t_1 + 1) = x_1$)

$$\begin{aligned}\beta_2 &= (x_1 - a_1/\mu)e^{\mu(t_1+1)} = (a_1/\mu)(-2 + e^{-\mu})e^{\mu}\frac{\mu h + a_1}{a_1} \\ &= (1 - 2e^{\mu}) \cdot \frac{\mu h + a_1}{\mu} = (1 - 2e^{\mu})(h + a_1/\mu).\end{aligned}$$

It follows as above for $\theta_\pm$ that the numbers $\vartheta_\pm = \vartheta_\pm(\delta)$ are independent of h, and given by the formulas from Proposition 3.1 with $\alpha := \alpha_2 = a_1/\mu$. On the interval $[t_2 + \vartheta_-, t_2 + \vartheta_+]$, the solutions x and $\tilde{x}$ coincide and pass through $[-\delta, \delta]$ a second time, but (obviously) in a different way. Since $\tilde{f}$ is already chosen, the analogue of condition (13) with α_2 instead of α will not hold, so that we do not obtain $\tilde{x}(t_2 + \vartheta_+ + 1) = x(t_2 + \vartheta_+ + 1)$. However, we have $\tilde{x}(t_2 + \vartheta_- + 1) = x(t_2 + \vartheta_- + 1)$, and for $t \in [t_2 + \vartheta_- + 1, t_2 + \vartheta_+ + 1]$ we have in analogy to (14)

$$\tilde{x}(t) - x(t) = \int_{t_2+\vartheta_-+1}^{t} e^{-\mu(t-s)} a_2 \cdot [\tilde{f}(x(s-1)) - f(x(s-1))]\, ds$$

(recall that $\tilde{x} = x$ on $[t_2 + \vartheta_-, t_2 + \vartheta_+]$), which gives the estimate $|\tilde{x}(t) - x(t)| \le (\vartheta_+ - \vartheta_-) \cdot 2a_2 \le C_2\delta$, with appropriate C_2. Assertions (b) and (c) follow with $C := \max\{C_1, C_2\}$.
Further, for $t = t_2 + \vartheta_+ + 1$ in particular, we obtain from Proposition 3.1 b):

$$\begin{aligned}&\tilde{x}(t_2 + \vartheta_+ + 1) - x(t_2 + \vartheta_+ + 1) \\ &= \int_{t_2+\vartheta_-+1}^{t_2+\vartheta_++1} e^{-\mu(t_2+\vartheta_++1-s)} a_2 \cdot [\tilde{f}(x(s-1)) - f(x(s-1))]\, ds \\ &= a_2 \cdot J(\alpha_2, \tilde{f} - f), \quad \text{where } \alpha_2 = a_1/\mu.\end{aligned}$$

Obviously, the last quantity, which we abbreviate by $R(\delta)$, is independent of the 'initial value' h and satisfies $|R(\delta)| \le C_2\delta \le C\delta$.

(Ad d): From (3) and (3.2) we see that for $t \in [t_2 + \vartheta_+ + 1, p_1 + p_2]$

$$\begin{aligned}\tilde{x}(t) - x(t) &= e^{-\mu[t-(t_2+\vartheta_++1)]} R(\delta) + \\ &+ \int_{t_2+\vartheta_++1}^{t} e^{-\mu(t-s)} a_2 \cdot [\tilde{f}(\tilde{x}(s-1)) - f(x(s-1))]\, ds,\end{aligned}$$

Now for h close to h_* and δ sufficiently small, both terms $\tilde{f}(\tilde{x}(s-1))$ and $f(x(s-1))$ in this integral equal -1, since the periodic solution x^{h_*} decreases from its maximal value at $t_2(h_*) + 1$ to the still positive value h_* on $[t_2(h_*), p_1 + p_2]$, and the difference $\tilde{x}(t) - x(t)$ equals the first summand in the above expression as long as also $\tilde{x}(s-1) \geq \delta$, which is true if $h_* + R(\delta) > \delta$ and $|h - h_*|$ is sufficiently small. We thus obtain the assertion of (d). □

3.3 Smoothing the Coefficient Function *a* to $\tilde{a}$

Proposition 3.3 *For sufficiently small $\rho > 0$, the step function a with the two positive values a_1, a_2 can be replaced by a function $\tilde{a} : \mathbb{R} \to \mathbb{R}$ which is of class C^∞, also has period $p_1 + p_2$, and the following properties:*

(a) $\tilde{a} = a = a_1$ on $[\rho, p_1]$, $\tilde{a} = a = a_2$ on $[p_1 + \rho, p_1 + p_2]$, hence also $\tilde{a} = a = a_2$ on $[-p_2 + \rho, 0]$.

(b) $0 < \tilde{a} \leq 2\max\{a_1, a_2\}$ on the 'transition intervals' $[0, \rho]$ and $[p_1, p_1 + \rho]$.

(c) For h close to h_*, the solutions $\tilde{\tilde{x}} = x^h(\cdot, \tilde{a}, \tilde{f})$ and $\tilde{x} = x^h(\cdot, a, \tilde{f})$ as in Proposition 3.2 satisfy

$$\tilde{\tilde{x}} = \tilde{x} \quad \text{on } [0, p_1 + p_2] \setminus ([0, \rho] \cup [p_1, p_1 + \rho]),$$

and on the 'exceptional' intervals $[0, \rho]$ and $[p_1, p_1 + \rho]$ one has

$$|\tilde{\tilde{x}} - \tilde{x}| \leq 2\max\{a_1, a_2\}\rho.$$

Proof We first replace a on an interval $[0, \rho]$ with $\rho > 0$ small by a C^∞ function $\hat{a}$ with $\hat{a}(0) = a_2, \hat{a}(\rho) = a_1$ (taking $\rho < t_1(h_*)$). The condition for the corresponding solution $\hat{x} = x^h(\cdot, \hat{a}, \tilde{f})$ to satisfy $\hat{x}(\rho) = \tilde{x}(\rho)$ is

$$\int_0^\rho \hat{a}(s)e^{-\mu(\rho - s)} \cdot (-1)\, ds = \int_0^\rho a(s)e^{-\mu(\rho - s)} \cdot (-1)\, ds, \text{ or}$$

$$\int_0^\rho \hat{a}(s)e^{-\mu(\rho - s)}\, ds = a_1 \int_0^\rho e^{-\mu(\rho - s)}\, ds.$$

This equality can be satisfied, along with the boundary conditions, by a function $\hat{a}$ taking values in $(0, a_2]$ if $a_1 < a_2$, and with values in $(0, 2a_1]$ if $a_2 < a_1$, so in any case with values in $(0, 2\max\{a_1, a_2\}]$. With such a choice, we obtain for $t \in [0, \rho]$

$$|\hat{x}(t) - x(t)| \leq 2\max\{a_1, a_2\} \int_0^t e^{-\mu(t - s)}\, ds \leq 2\max\{a_1, a_2\}\rho,$$

and $\hat{x}(\rho) = \tilde{x}(\rho)$. For small enough ρ we conclude that $\tilde{f}(\hat{x}(t)) = \tilde{f}(\tilde{x}(t)) = -1$ for $t \in [0, \rho]$, implying $\hat{x} = \tilde{x}$ on $[\rho, p_1 + p_2]$, where $\hat{a} = a$.

In a similar second step we modify $\hat{a}$ to $\tilde{a}$ on the interval $[p_1, p_1 + \rho]$ (choosing $\rho > 0$ such that $p_1 + \rho < t_2(h_*) + 1$) to a C^∞ function $\tilde{a}$ with period $p_1 + p_2$, with $\tilde{a}(p_1) = a_1, \tilde{a}(p_1 + \rho) = a_2$, and $\tilde{a} = \hat{a}$ on $[0, p_1 + p_2] \setminus [p_1, p_1 + \rho]$. Again, this can be done with $\tilde{a}$ taking values in $(0, 2\max\{a_1, a_2\}]$ and such that the solutions $\hat{x} = x^h(\cdot, \hat{a}, \tilde{f})$ (as above) and $\tilde{\tilde{x}} := x^h(\cdot, \tilde{a}, \tilde{f})$ with the same value h close to h_* at time zero satisfy $\hat{x}(p_1 + \rho) = \tilde{\tilde{x}}(p_1 + \rho)$. As above, then for ρ small enough (so that $\tilde{\tilde{x}} \geq \delta$ on $[p_1, p_1 + \delta]$), this gives $\tilde{\tilde{x}} = \hat{x}$ on $[p_1 + \rho, p_1 + p_2]$. □

Corollary 3.4 *For the equation $x'(t) = -\mu x(t) + \tilde{a}(t)\tilde{f}(x(t-1))$, with $\tilde{a}$ and $\tilde{f}$ as above, and for h close to h_*, the map $\tilde{F} : h = x(0) \mapsto x(p_1 + p_2) = \tilde{F}(h)$ is given by*

$$\tilde{F}(h) = F(h) + \eta(h, \delta)R(\delta) = mh + b + \eta(h, \delta)R(\delta), \quad \textit{with}$$

$$\eta(h, \delta) := \exp[\mu(1 - (p_1 + p_2))](2e^{\mu} - 1)\frac{a_1 + \mu h}{a_1 - \mu\delta},$$

with R from Proposition 3.2, and m, b as in Sect. 2.1.

Proof The formulas for t_1, x_1 and t_2 from Section 2.1 give $x_1 = \frac{a_1}{\mu}(e^{-\mu} - 1)$ and

$$\begin{aligned} t_2 = t_2(h) &= t_1(h) + 1 + \frac{1}{\mu}\ln[\frac{a_1 - \mu\frac{a_1}{\mu}(e^{-\mu} - 1)}{a_1}] = t_1(h) + 1 + \frac{1}{\mu}\ln(2 - e^{-\mu}) \\ &= 1 + \frac{1}{\mu}[\ln(\frac{a_1 + \mu h}{a_1}) + \ln(2 - e^{-\mu})] \\ &= 1 + \frac{1}{\mu}\ln[(2 - e^{-\mu})\frac{a_1 + \mu h}{a_1}]. \end{aligned}$$

Thus $e^{\mu t_2(h)} = e^{\mu}(2 - e^{-\mu})\frac{a_1 + \mu h}{a_1} = (2e^{\mu} - 1)\frac{a_1 + \mu h}{a_1}$. Further note that from the construction of $\vartheta_+(\delta)$ and Proposition 3.1,

$$e^{\mu\vartheta_+(\delta)} = \frac{1}{1 - \delta/\alpha_2} = \frac{1}{1 - \mu\delta/a_1} = \frac{a_1}{a_1 - \mu\delta}.$$

Combining part (c) of Proposition 3.3 and part (d) of Proposition 3.2 we see that

$$\begin{aligned} \tilde{F}(h) = \tilde{\tilde{x}}(p_1 + p_2) &= x(p_1 + p_2) + \exp\{-\mu[p_1 + p_2 - (t_2(h) + \vartheta_+(\delta) + 1)]\}R(\delta) \\ &= F(h) + \exp[\mu(1 - (p_1 + p_2))]e^{\mu t_2(h)}e^{\mu\vartheta_+(\delta)}R(\delta) \\ &= F(h) + \exp[\mu(1 - (p_1 + p_2))](2e^{\mu} - 1)\frac{a_1 + \mu h}{a_1}\frac{a_1}{a_1 - \mu\delta}R(\delta) \\ &= F(h) + \exp[\mu(1 - (p_1 + p_2))](2e^{\mu} - 1)\frac{a_1 + \mu h}{a_1 - \mu\delta}R(\delta) \\ &= F(h) + \eta(h, \delta)R(\delta). \end{aligned}$$

□

With the smoothing as in Corollary 3.4, we obtain the following theorem for the equation

$$(\tilde{a}, \tilde{f}) \qquad x'(t) = -\mu x(t) + \tilde{a}(t)\tilde{f}(x(t-1))$$

Theorem 3.5 *For a smoothed equation $(\tilde{a}, \tilde{f})$ as in Corollary 3.4 and $\delta > 0$ small enough, the map $\tilde{F}$ a has a unique fixed point $\tilde{h}$ close to h_*, with*

$$\tilde{h} \to h_* \text{ and } \tilde{F}'(\tilde{h}) \to F'(h_*) = m \text{ as } \delta \to 0.$$

To this fixed point corresponds a periodic solution $\tilde{p} : \mathbb{R} \to \mathbb{R}$ of equation $(\tilde{a}, \tilde{f})$ with (minimal) period $p_1 + p_2$.

Proof The proof follows from the construction and from the fact that in a neighborhood of h_*, the map $\tilde{F}$ converges to the affine linear map F in the C^1-norm, as one sees from the description of their difference in Corollary 3.4 and the property $R(h) \to 0$ $(\delta \to 0)$ from Proposition 3.2. □

Intuitively, the conclusion of Theorem 3.5 is expected due to the continuous dependence of the solutions of Eq. (3) on the parameter $\delta > 0$ when one replaces discontinuous functions f_0 and A_0 of Sect. 2 by 'smoothing' them in a small δ-neighborhood of their discontinuity points only. The smooth solutions of the modified equation $(\tilde{a}, \tilde{f})$ are uniformly close on compact intervals to the non-differentiable ones from Sect. 2 (see Figs. 1 and 2 for the latter). In fact they are made up of the same piece-wise exponential segments everywhere except in small δ-size neighborhoods of the previously non-differentiability points, where they are now of at least C^1-smoothness (i.e., points $p_1, t_1 + 1$, and $p_1 + p_2$, where the 'corners' are smoothed to C^1-class curves; see Fig. 2). This theorem shows that the periodic dynamics and their stability persist when $\delta > 0$ is small enough. See also the remarks on C^1- small perturbations of the smoothed equation directly after Theorem 3.6 below.

3.4 Computation of Spectral Properties of the Smoothed Period Map

Consider a smoothed equation $(\tilde{a}, \tilde{f})$ as in Theorem 3.5 with δ small enough, so that the corresponding periodic solution $\tilde{p}$ exists. (We generally do not denote the dependence of objects on δ or other details of the smoothing.) Note that the initial state $\tilde{\varphi} := \tilde{p}_0$ of this periodic solution is given by $p_0 = \tilde{\tilde{x}}_{p_1+p_2}$, where as above, $\tilde{\tilde{x}} = x^{\tilde{h}}(\cdot, \tilde{a}, \tilde{f})$, and one can take the constant function equal to $\tilde{h}$ as initial state for $\tilde{\tilde{x}}$. We briefly write C^0 for the space $C^0([-1, 0], \mathbb{R})$ with the norm given by

$$||\varphi||_\infty = \max\Big\{|\varphi(t)| \;\Big|\; t \in [-1, 0]\Big\},$$

and define the evaluation map $\mathrm{ev}_0 : \varphi \in C^0([-1, 0], \mathbb{R}) \to \mathbb{R}$, $\mathrm{ev}_0(\varphi) := \varphi(0)$. Set

$$D^+ := \left\{ \varphi \in C^0 \mid \varphi([-1, 0]) \subset [\delta, \infty) \right\}.$$

For initial functions $\varphi \in D^+$, (and with the 'starting time' $t_0 = 0$) the values of the corresponding solution x^φ on $[0, 1]$ (the first interval of delay length) depend only on $h := \varphi(0) = \mathrm{ev}_0(\varphi)$, and hence the same is true for the solution on all of $[0, \infty)$. From Theorem 3.5, the fixed point $\tilde{h} > 0$ of $\tilde{F}$ (close to h_*) exists.

Set $T := p_1 + p_2$, and define the period map $P(T, 0) : C^0 \ni \varphi \mapsto x^\varphi(\cdot, \tilde{a}, \tilde{f})_T \in C^0$, which is the forward evolution operator for the nonautonomous equation from starting time 0 to the period T. The fact that $\tilde{a}$ and $\tilde{f}$ are of class C^1 (for $\tilde{a}$, continuity would be sufficient) implies that P is of class C^1 w.r. to the norm $|| \, ||_\infty$ (compare, e.g., Theorem 4.1, p. 46 in Hale 1977). Due to the nonautonomous character of the equation, the starting time is also important here, not only the time difference T. Analogous time evolution operators $P(t, s)$ can be defined for all $t, s \in \mathbb{R}$, which then have the cocycle property $P(t, s) \circ P(s, u) = P(t, u)$ (if $t \geq s \geq u$). T-periodicity of the equation implies that

$$P(nT, 0) = [P(T, 0)]^n \; (n \in \mathbb{N}_0). \tag{15}$$

We mostly write P for $P(T, 0)$ from now on. It follows from (15) that fixed points of P give periodic solutions of the equation, and that iteration of P describes the solution behavior also for non-periodic solutions. In particular,

$$P(\tilde{\varphi}) = \tilde{\varphi}.$$

For an appropriate open interval $I \subset \mathbb{R}$ containing $\tilde{h}$, and the set $\tilde{D}^+ := \left\{ \varphi \in D^+ \mid \varphi(0) \in I \right\}$ we have a commutative diagram

$$\begin{array}{ccc} \tilde{D}^+ & \xrightarrow{\;P\;} & D^+ \\ {\scriptstyle \mathrm{ev}_0}\downarrow & & \downarrow{\scriptstyle \mathrm{ev}_0} \\ I & \xrightarrow{\;\tilde{F}\;} & \mathbb{R} \end{array}$$

Now, for $h \in I$, consider the constant functions $h \cdot \mathbf{1} \in C^0$, where $\mathbf{1} \in C^0$ is the constant function with value 1. Define $\Gamma : I \ni h \mapsto P(h \cdot \mathbf{1}) \in C$; then $\Gamma : I \to C^0$ is a C^1-curve (we do not denote the dependence of Γ on δ.) The fact that forward solutions depend only on the value of the initial segment at $t = 0$ implies that for a whole neighborhood U in C^0 of the initial segment $\tilde{\varphi}$ of the periodic solution (the fixed point of P) the following inclusion is valid

$$P(U) \subset \mathrm{image}(\Gamma),$$

and we have

$$\Gamma(\tilde{h}) = P[\tilde{h} \cdot \mathbf{1}] = P(\tilde{\varphi}) = \tilde{\varphi}.$$

We know (assuming δ small enough) that $\lambda := \tilde{F}'(\tilde{h}) \approx m$ is not zero. Note that the map $I \ni h \mapsto \mathrm{ev}_0(h \cdot \mathbf{1})$ is the identity on I, and hence we have

$$\tilde{F}'(\tilde{h}) = \frac{d}{dh}\Big|_{\tilde{h}}\{\tilde{F} \circ ev_0 \circ [h \mapsto h \cdot \mathbf{1}]\}.$$

Using the above diagram we obtain

$$\begin{aligned} 0 \neq \lambda = \tilde{F}'(\tilde{h}) &= \frac{d}{dh}\Big|_{h = \tilde{h}}\{\tilde{F} \circ \mathrm{ev}_0 \circ [h \mapsto h \cdot \mathbf{1}]\} \\ &= \frac{d}{dh}\Big|_{\tilde{h}}\{\mathrm{ev}_0 \circ P \circ [h \mapsto h \cdot \mathbf{1}]\} \\ &= \frac{\mathrm{d}}{\mathrm{d}h}\Big|_{h = \tilde{h}}[\mathrm{ev}_0 \circ \Gamma] = \mathrm{ev}_0[\Gamma'(\tilde{h})]. \end{aligned}$$

Thus, the tangent vector $v := \Gamma'(\tilde{h}) \in C^0$ is nonzero, and even $v(0) \neq 0$. It follows that there exist an open interval I^* around $\tilde{h}$ such that $W^* := \mathrm{image}(\Gamma_{|I^*})$ is a one-dimensional C^1-submanifold of C^0, containing $\tilde{\varphi}$. A neighborhood of $\tilde{\varphi}$ in C^0 is mapped into W^* by P. Therefore, with the one-dimensional tangent space $\mathcal{T} := T_{\tilde{\varphi}}W^* = \mathbb{R} \cdot \Gamma'(\tilde{\varphi}) = \mathbb{R} \cdot v$ and the derivative $M := DP(\tilde{\varphi})$ one has $\mathrm{image}(M) \subset \mathcal{T}$. We conclude that the kernel $\mathcal{N}$ of M is a closed subspace of C^0 with codimension at most one, and $M\mathcal{T} \subset \mathcal{T}$, so that v is an eigenvector of M. (We show next that the corresponding eigenvalue equals λ, which comes as no surprise.) "Differentiating" the above diagram at $\tilde{\varphi}$ into the direction of v, we obtain

$$\begin{aligned} (Mv)(0) &= \mathrm{ev}_0(M(v)) = \mathrm{ev}_0(DP(\tilde{\varphi})(v)) \\ &= \tilde{F}'(\mathrm{ev}_0(\tilde{\varphi})) \cdot \mathrm{ev}_0(v) = \tilde{F}'(\tilde{h}) \cdot \mathrm{ev}_0(v) = \lambda \cdot v(0). \end{aligned}$$

Since $v(0) \neq 0$, and we know already that v is an eigenvector of M, the associated eigenvalue equals λ. Summing up, we obtain the following result:

Theorem 3.6 *Assume that for the original discontinuous Eq. (3) the map F defined by $F(h) = mh + b$ as in Sect. 2.1 has a fixed point $h_* > 0$. Then, for smoothed equations $(\tilde{a}, \tilde{f})$ as above, if $\delta > 0$ is small enough, there exist a unique periodic solution $\tilde{p}$ with period $T = p_1 + p_2$ and initial state $P_0 = \tilde{\varphi}$, where $\varphi(0) = \tilde{h}$ is the unique fixed point of the associated one-dimensional map $\tilde{F}$. The corresponding period map $P = P(T, 0)$ with $P(\tilde{\varphi}) = \tilde{\varphi}$ maps a neighborhood of $\tilde{\varphi}$ in C^0 into a one-dimensional submanifold of C^0. For its derivative $M = DP(\tilde{\varphi})$, we have the $M-$invariant decomposition*

$$C^0 = \mathcal{N} \oplus \mathcal{T}$$

corresponding to the eigenvalues 0 *and* λ *of* M*. Further,* $\lambda \to m$ *as* $\delta \to 0$*, so, in particular, the periodic solution* $\tilde{p}$ *is stable for equation* $(\tilde{a}, \tilde{f})$ *if* $|m| < 1$*, and* δ *is small enough.*

Remarks on C^1-small perturbations of the smoothed equation. If $m \neq 1$, it also follows from standard C^1-perturbation theory for hyperbolic fixed points that, for equations with data C^1-close to $\tilde{a}$ and $\tilde{f}$, there will exist a unique nearby periodic solution (the initial value of which is a fixed point of the perturbed period map) with analogous stability properties. We do not prove this in detail; one would need a non-autonomous (here: periodic) statement of the type 'nearby equations generate C^1– close period maps, with nearby fixed points', i.e., a non-autonomous version of Theorem 1.8, p. 20 from Lani-Wayda (1995a). The robustness of the fixed point of the period map P (and of the spectral properties of $M = DP(\varphi)$) are then a very special case of perturbation arguments for hyperbolic sets for non-invertible mappings, e.g., Theorem 6.6 from Lani-Wayda (1995b).

Remarks on the double period solutions. Finally, we sketch how analogous considerations apply to the solutions with double period from Sect. 2.2. In this case, for the smoothed equation the period map P corresponds, on positive initial functions, to a one-dimensional map $\tilde{\Phi}_1$ which is C^1-close to Φ_1. One has to prove then that P corresponds, on negative initial functions, to a one-dimensional map $\tilde{\Phi}_2$ which is C^1-close to Φ_2. (This is not immediate from the first part, since the property that solutions of smooth and non-smooth equations coincide except for small intervals gets lost after the second crossing of zero.) Then the second iterate $P^2 = P(2T, T) \circ P(T, 0) = P(T, 0) \circ P(T, 0)$ corresponds to $\tilde{\Phi}_2 \circ \tilde{\Phi}_1$, which is C^1 close to $\Phi_2 \circ \Phi_1$. Now, for general h, one has in view of (8):

$$(\Phi_2 \circ \Phi_1)(h) = \Phi_2(kh + d) = -\Phi_1(-(kh + d)) = -[k \cdot (-(kh + d)) + d] = k^2 h + kd - d.$$

Note that h_* is a fixed point of $\Phi_2 \circ \Phi_1$, and the derivative there equals k^2. Then existence and stability properties of this fixed point transfer to the map P^2 for the smoothed equation as described for the single period solutions in Sect. 3.

The above considerations are similar to the ones in connection with the dynamics generated by autonomous delay equations with smoothed step functions, as studied in Walther (1981b), Hale and Lin (1986), and Lani-Wayda (1995a). In particular, the argument here leading to Theorem 3.6 is similar to the one from Lani-Wayda (1995a), pp. 31–34.

Acknowledgements This collaborative research project was initiated during A.I.'s visit and stay at the Justus-Liebig Universität in Giessen, Germany, under the support from the Alexander von Humboldt Stiftung (June-August 2023). It was further advanced at the mathematical research institute MATRIX in Australia during the workshop "Delay Differential Equations and Their Applications" held in December 2023 (https://www.matrix-inst.org.au/events/delay-differential-equations-and-their-applications/).

References

Ai, S.: An improved result on bounded/periodic solutions for some scalar delay differential equations. Discret. Contin. Dyn. Syst. Ser. S **17**(4), 1639–1647 (2024)

Amster, P., Idels, L.: Periodic solutions in general scalar non-autonomous models with delays. Nonlinear Differ. Equ. Appl. **20**(5), 1577–1596 (2013)

An der Heiden, U., Walther, H.-O.: Existence of chaos in control systems with delayed feedback. J. Differ. Equ. **47**, 273–293 (1983)

An der Heiden, U., Mackey, M.C.: The dynamics of production and destruction: analytic insight into complex behavior. J. Math. Biol. **16**, 75–101 (1982)

Diekmann, O., van Gils, S., Verdyn Lunel, S., Walther, H.-O.: Delay Equations: Complex, Functional, and Nonlinear Analysis. Applied Mathematical Sciences, vol. 110. Springer (1995)

Erneux, T.: Applied Delay Differential Equations. Ser.: Surveys and Tutorials in the Applied Mathematical Sciences, vol. 3. Springer (2009)

Faria, T.: Periodic solutions for a non-monotone family of delayed differential equations with applications to Nicholson systems. J. Differ. Equ. **263**(1), 509–533 (2017)

Friedman, H.I., Wu, J.: Periodic solutions of single-species models with periodic delay. SIAM J. Math. Anal. **23**(3), 689–701 (1992)

Glass, L., Mackey, M.C.: From Clocks to Chaos. The Rhythms of Life, Princeton University Press (1988)

Gurney, W.S.C., Blythe, S.P., Nisbet, R.M.: Nicholson's blowflies revisited. Nature **287**, 17–21 (1980)

Hale, J.K., Verduyn Lunel, S.M.: Introduction to Functional Differential Equations. Springer Applied Mathematical Sciences, vol. 99 (1993)

Hale, J.K.: Theory of Functional Differential Equations. Springer, New York (1977)

Hale, J.K., Lin, X.B.: Examples of transverse homoclinic orbits in delay equations. Nonlinear Anal. **10**, 693–709 (1986)

Ivanov, A.F., Sharkovsky, A.N.: Oscillations in singularly perturbed delay equations. In: Dynamics Reported (New Series) vol. 1, pp. 165–224. Springer (1992)

Ivanov, A., Shelyag, S.: Explicit periodic solutions in a delay differential equation. In: Addressing Modern Challenges in the Mathematical, Statistical, and Computational Sciences. Springer Proceedings in Mathematics and Statistics, Vol. 494, pp. 459-466. Springer Nature Switzerland AG (2025)

Ivanov, A., Shelyag, S.: Periodic solutions in a simple delay differential equation. Math. Comput. Appl. **29**, 36 (2024)

Ivanov, A., Shelyag, S.: Stable periodic solutions in scalar periodic differential delay equations. Archivum Mathematicum **59**(1), 69–76 (2023). Proceedings of International Conference EquaDiff 15, 11–15 July 2022, Masaryk University, Brno, Czech Republic

Jiang, D., Wei, J., Zhang, B.: Positive periodic solutions of functional differential equations and population models. Electron. J. Differ. Equ. **2002**(71), 1–13 (2002)

Kuang, Y.: Delay Differential Equations with Applications in Population Dynamics. Mathematics in Science and Engineering, vol. 191, 398 pp. Academic Press (1993)

Lani-Wayda, B.: Hyperbolic sets, shadowing and persistence for noninvertible mappings in Banach spaces. Pitman Research Notes in Mathematics Series, vol. 334. Longman, Essex, and Wiley, New York (1995)

Lani-Wayda, B.: Persistence of Poincaré mappings in functional differential equations (with application to structural stability of complicated behavior). J. Dynam. Differ. Equ. **7**(1), 1–71 (1995)

Li, Y., Kuang, Y.: Periodic solutions in periodic state-dependent delay equations and population models. Proc. Amer. Math. Soc. **130**, 1345–1353 (2001)

Liu, Y., Yang, P., Weigao, G.: Periodic solutions of higher-order delay differential equations. Nonlinear Anal. **63**, 136–152 (2005)

Mackey, M.C., Glass, L.: Oscillation and chaos in physiological control systems. Science **197**, 287–289 (1977)

Perez, J.F., Malta, C.P., Coutinho, F.A.B.: Qualitative analysis of oscillations in isolated populations of flies. J. Theoret. Biol. **71**, 505–514 (1978)
Peters, H.: Chaotic behavior of nonlinear differential-delay equations. Nonlinear Anal. **7**(12), 1315–1334 (1983)
Smith, H.: An Introduction to Delay Differential Equations with Applications to the Life Sciences. Texts in Applied Mathematics, vol. 57 (2011). Springer
Walther, H.-O.: Density of slowly oscillating solutions of $\dot{x}(t) = -f(x(t-1))$. J. Math. Anal. Appl. **79**(12), 127–140 (1981)
Walther, H.-O.: Homoclinic solution and chaos in $\dot{x}(t) = f(x(t-1))$. Nonlinear Anal. **5**(7), 775–788 (1981)
Wazewska-Czyzewska, M., Lasota, A.: Mathematical models of the red cell system. Matematyka Stosowana **6**, 25–40 (1976). (in Polish)

Adaptive Mesh Construction for the Numerical Solution of Stochastic Differential Equations with Markovian Switching

Cónall Kelly and Kate O'Donovan

Abstract We demonstrate an approach to the numerical solution of nonlinear stochastic differential equations with Markovian switching. Such equations describe the stochastic dynamics of processes where the drift and diffusion coefficients are subject to random state changes according to a Markov chain with finite state space. We propose a variant of the Jump Adapted-Adaptive approach introduced by Lord and Sun (2024) to construct nonuniform meshes for explicit numerical schemes that adjust timesteps locally to rapid changes in the numerical solution and which also incorporate the switching times of an underlying Markov chain as meshpoints. It is shown that a hybrid scheme using such a mesh that combines an efficient explicit method (to be used frequently) and a potentially inefficient backstop method (to be used occasionally) will display strong convergence in mean-square of order δ if both methods satisfy a mean-square consistency condition of the same order in the absence of switching. We demonstrate the construction of an order $\delta = 1$ method of this type and apply it to generate empirical distributions of a nonlinear SDE model of telomere length in DNA replication.

Keywords Numerical solution · Adaptive mesh construction · Stochastic differential equation · Jump Adapted-Adaptive approach · Markovian switching · Nonlinear SDE model of telomere length in DNA replication

1 Introduction

We examine a method of constructing adaptive meshes for the efficient numerical solution of d-dimensional system of stochastic differential equations with Markovian switching (SDEwMS) of the form

C. Kelly (✉) · K. O'Donovan
School of Mathematical Sciences, University College Cork, Cork, Ireland
e-mail: conall.kelly@ucc.ie

K. O'Donovan
e-mail: 116370906@umail.ucc.ie

E. Braverman and A. Swishchuk (eds.), *Delay and Stochastic Differential Equations*, ICIAM2023 Springer Series 12, https://doi.org/10.1007/978-981-95-7310-3_3

$$\begin{aligned} dX(t) &= f(X(t), r(t))dt + g(X(t), r(t))dW(t), \quad 0 \le t \le T, \\ X(0) &= x_0 \in \mathbb{R}^d. \end{aligned} \tag{1}$$

In (1), the drift and diffusion coefficients f and g take the form

$$f : \mathbb{R}^d \times \mathbb{S} \to \mathbb{R}^d, \quad \text{and} \quad g : \mathbb{R}^d \times \mathbb{S} \to \mathbb{R}^{d\times m}.$$

The stochastic dynamics of a process governed by (1) are influenced by two independent sources of noise. The process r dictates changes occurring at random intervals in the functional form of f and g, and is described by a scalar continuous time Markov chain taking values in the set $\mathbb{S} = \{1, 2, .., L\}$. The process W is an m-dimensional Brownian motion and captures the effect of a diffusion process on trajectories. We characterise both mathematically in Sect. 2.

We are motivated by the following example where a scalar SDE is used to model the shortening over time of telomere length in DNA replication in jackdaws:

$$dL(t) = -\left(c + aL(t)^2\right) dt + \sqrt{\frac{1}{3}aL(t)^3 dW(t)}, \tag{2}$$

where W is a scalar standard Brownian motion. As presented in Grasman et al. (2011), the model arises as the Langevin equation associated with the Fokker-Planck PDE

$$\frac{\partial p}{\partial t} = \frac{\partial}{\partial l}\left\{\left(c + \frac{1}{2}al^2\right) p\right\} + \frac{1}{2}\frac{\partial^2}{\partial l^2}\left\{\left(\frac{1}{3}al^3\right) p\right\},$$

which governs the probability density function $p(t, l)$ for the telomere length l at time t. By simulating an ensemble of trajectories of (2) we can produce a density histogram of lengths $l = L$ at a fixed time $t = T$ which serves to numerically approximate $p(T, L)$. The parameters c and a control the underlying decay rate of the length and the intensity at which random breaks occur in the telomere respectively.

The general form of the SDE (1) without Markovian switching, can be written

$$\begin{aligned} dX(t) &= f(X(t))dt + g(X(t))dW(t), \quad 0 \le t \le T, \\ X(0) &= x_0 \in \mathbb{R}^d. \end{aligned} \tag{3}$$

We can see that the telomere length SDE (2) is a special case of (3) with $d = m = 1$, $f(x) = -(c + ax^2)$, and $g(x) = (ax^3/3)^{1/2}$.

Suppose we wish to solve (3) numerically over an arbitrary mesh $\{t_n\} = \{0 = t_0, t_1, \ldots, t_{N-1}, t_N = T\}$. Over any step $[t_n, t_{n+1}]$, solutions of (3) satisfy the integral equation

$$X(s) = X(t_n) + \int_{t_n}^{s} f(X(r))dr + \int_{t_n}^{s} g(X(r))dW(r), \quad s \in [t_n, t_{n+1}]. \tag{4}$$

By fixing the solution at the beginning of the interval $[t_n, t_{n+1}]$, we achieve a continuous form of the explicit Euler-Maruyama approximation

$$\bar{X}(s) = X_n + \int_{t_n}^{s} f\left(\bar{X}(t_n)\right) dr + \int_{t_n}^{s} g\left(\bar{X}(t_n)\right) dW(r), \quad s \in [t_n, t_{n+1}]. \tag{5}$$

Setting $X_n := \bar{X}(t_n)$, we can write the scheme in its computational form as

$$X_{n+1} = X_n + h_{n+1} f(X_n) + g(X_n)\Delta W_{n+1}, \quad n = 0, \ldots, N-1, \tag{6}$$

where $h_{n+1} = t_{n+1} - t_n$ and $\Delta W_{n+1} = W(t_{n+1}) - W(t_n)$. Suppose there exists a constant maximum stepsize $h_{\max} \in (0, \infty)$ such that $h_n \leq h_{\max}$ for all $n = 0, \ldots, N$. We say that solutions of (5) (or equivalently (6)) converge strongly to solutions of (3) in L_p with order γ if and only if

$$\max_{t \in [0,T]} \left(\mathbb{E}\left[\|X(t) - \bar{X}(t)\|^p\right]\right)^{1/p} \leq C h_{\max}^{\gamma}.$$

If $p = 2$ then we say that (5) converges strongly in L_2, in mean-square. It is the case that the Euler-Maruyama method (5) converges in mean-square to (3) with order $\gamma = 1/2$ when f, g are globally Lipschitz and $h_n \equiv h$.

Hutzenthaler et al. (2011) confirmed that solutions of (6) with constant stepsize $h_n \equiv h$ fail to converge strongly in the strong sense to solutions of (3) if either f or g grow superlinearly in norm. Indeed this is the case for the telomere model (2). To address this issue, many novel explicit methods have been proposed that converge strongly even for highly nonlinear drift and diffusion coefficients; these include tamed, projected, truncated, and adaptive variants of both Euler-Maruyama and Milstein methods (see, for example Beyn et al. 2017; Guo et al. 2018; Hutzenthaler et al. 2012; Mao 2015; Wang and Gan 2013). In particular we highlight the adaptive timestepping approach that was introduced for the Euler-Maruyama method in Kelly and Lord (2018) and extended to the Milstein method in Kelly et al. (2023), where it was used to construct a numerical method converging strongly with order $\gamma = 1$ and to generate trajectories of the telomere SDE (2).

We wish to extend this capability to models where the coefficients f and g undergo Markovian switching. For example in the case of the telomere SDE (2) it may be that each of the model parameters in the pair (c, a) undergo transition between two distinct values according to a Markov process. Alternatively one can view Markovian switching in the model parameters as a form of stress testing to investigate the robustness of sample distributions to random changes in these parameters.

We take as our template the approach to mesh construction in Kelly et al. (2025), where it was shown that an adaptive numerical scheme that satisfies a particular mean-square consistency bound can be extended to ensure strong convergence in mean-square for SDEs with jump perturbations (SJDEs) as long as the jump times are incorporated into the adaptive mesh in a particular way. In such a case the mesh is adapting both to the evolving behaviour of the approximate trajectory and directly to jumps when they occur. We refer to these as Jump Adapted-Adaptive Methods (JA-AMs).

In this article, we show that an analogue of the type of mesh used for JA-AMs also ensures convergence of explicit numerical schemes satisfying a similar mean-square consistency condition to SDEwMS governed by (1), as long as we include the times at which the Markov chain r switches between distinct states. We prove strong convergence of such schemes in mean-square, noting that the major challenge of this analysis is dealing with the fact that the number of Markov transitions on any interval is finite but unbounded, and we handle it in the same manner as in the jump case (Kelly et al. 2025). Finally we demonstrate the practical application of this approach on a variant of SDE model of telomere shortening (2) with stochastic dynamics governed by coefficients that undergo transitions at random times.

In Sect. 2 we provide a complete mathematical description of the setting, including of W and r, and provide conditions on f and g for which unique solutions of (1) exist on $[0, T]$ for $T < \infty$. In Sect. 3 we describe how an adaptive mesh can be constructed that adapts both to the nonlinear functional response affecting the numerical solutions of (1) and which incorporate the switching times of r. We describe the general form of our adaptive scheme and present a version of it based upon the Milstein method. In Sect. 4 we state and prove our man strong convergence theorem, which in particular confirms that the Milstein scheme applied over this class of adaptive meshes will converge strongly in L_2 with order $\delta = 1$ to solutions of (1). Finally in Sect. 5 we describe the practical implementation of the scheme and demonstrate its use to generate sample histograms of telomere length in base pairs using a nonlinear SDEwMS model.

2 Mathematical Preliminaries

In this section, we mathematically characterise the random processes W and r influencing the dynamics of (1), then specify additional constraints on the coefficients f and g that are necessary for our main result.

2.1 *Characterisation of Random Inputs*

First, let $(\Omega, \mathcal{F}, (\mathcal{F}_t)_{t\geq 0}, \mathbb{P})$ be a complete probability space with a filtration $(\mathcal{F}_t)_{t\geq 0}$ which is right continuous and where $\mathcal{F}_0$ contains all $\mathbb{P}$-null sets. The process $W(t) = (W(t_1), ..., W(t_m))^T$ is an m-dimensional Brownian motion defined on this probability space such that each W_i is independent of W_j for $i \neq j$, and each $(W_i)_{t\in[0,T]}$ is a scalar stochastic process with $W_i(0) = 0$, a.s. continuous trajectories, and stationary increments that are independent on non-overlapping intervals.

Let $r(t), t \geq 0$ be a right-continuous Markov chain on the same probability space, taking values in $\mathbb{S}$ with generator $\Gamma = (\gamma_{ij})_{L\times L}$ given by

$$\mathbb{P}[r(t+\Delta) = j | r(t) = i] = \begin{cases} \gamma_{ij}\Delta + o(\Delta), & \text{if } i \neq j; \\ 1 + \gamma_{ii}\Delta + o(\Delta), & \text{if } i = j; \end{cases} \tag{7}$$

where $\Delta > 0$ and $o(\Delta)$ are higher order terms of Δ. Here $\gamma_{ij} \geq 0$ is the transition rate from i to j if $i \neq j$ while

$$\gamma_{ii} = -\sum_{j\neq i} \gamma_{ij}. \tag{8}$$

We assume that the standard Brownian Motion $W(\cdot)$ and the Markov chain $r(\cdot)$ are both $\mathcal{F}_t$-adapted, and for the latter the initial value $r(0) = r_0$ is an $\mathcal{F}_0$-measurable random variable taking values in $\mathbb{S}$.

The generator matrix Γ for a state space $\mathbb{S} = \{1, ..., L\}$ is

$$\Gamma = \begin{bmatrix} \gamma_{11} & \gamma_{12} & \cdots & \gamma_{1L} \\ \gamma_{21} & \gamma_{22} & \cdots & \gamma_{2L} \\ & & \cdot & \\ & & \cdot & \\ \gamma_{L1} & \gamma_{L2} & \cdots & \gamma_{LL} \end{bmatrix}. \tag{9}$$

From (8) we see that $\sum_{j=i}^{L} \gamma_{ij} = 1$ for all $i = 1, ..., L$. Denote the kth time at which the Markov chain switches between distinct states as τ_k. Then $(\tau_i)_{i\geq 0}$ is an increasing sequence of non-negative random variables which can be written

$$(\tau_i)_{i\geq 0} = \sum_{k=0}^{i} \pi_k, \tag{10}$$

where π_k denotes the waiting time between the kth and $(k+1)^{th}$ such transitions. In practice, as long as we keep track of the current state $i \in \mathbb{S}$ of the Markov chain, these waiting times may be computed as the holding times associated with each i, and these are modelled as exponentially distributed random variables with state-dependent rates λ_i: see Sect. 5.

Associated with the Markov chain $r(t)$ is a process $(\bar{N}_t)_{t\geq 0}$ that counts the number of switches that take place over the interval $[0, T]$ for all $t \geq 0$, written

$$\bar{N}_t := \#\{i \geq 1, \tau_i \in [0, T]\}. \tag{11}$$

Since the state space $\mathbb{S}$ is a finite set, $\bar{N}_t$ is almost surely (a.s) finite on any interval $[0, T]$: see for example (Norris 1997, Theorem 2.7.1). However $\bar{N}_t(\omega)$ is not uniformly bounded from above across the sample space Ω for any $t \in [0, T]$, and this presents a significant challenge in our main convergence result. A key to overcoming this challenge is the following.

Assumption 1 We suppose that $(\mathcal{F}_t)_{t\geq 0}$ can be decomposed such that $\mathcal{F}_t = \sigma(\mathcal{G}_t \cup \mathcal{H}_T)$ for all $t \geq 0$, where W is adapted to the filtration $(\mathcal{G}_t)_{t\geq 0}$, r is adapted to the filtration $(\mathcal{H}_t)_{t\geq 0}$, and moreover the σ-algebra $\mathcal{G}_t$ is independent of $\mathcal{H}_T$ for all $t \in [0, T]$.

Assumption 1 holds if $r(\cdot)$ is independent of $W(\cdot)$, and in practice it allows us to simulate trajectories of (1) by precomputing trajectories of r and incorporating switching times into an adaptive mesh.

2.2 *Minimal Conditions on the Drift and Diffusion Coefficients f and g*

We identify conditions on f and g that ensure the existence of unique solutions over each interval $[0, T]$. We require that both satisfy a local Lipschitz condition:

Assumption 2 For every integer $k \geq 1$, there exists a positive constant h_k such that, for all $i \in \mathbb{S}$ and those $x, y \in \mathbb{R}^n$ with $|x| \vee |y| \leq k$

$$\|f(x,i) - f(y,i)\|^2 \vee \|g(x,i) - g(y,i)\|^2 \leq \bar{h}_k \|x - y\|^2. \tag{12}$$

Consider also the following linear growth condition:

Assumption 3 f and g satisfy a linear growth bound if for all $(x, i) \in \mathbb{R}^d \times \mathbb{S}$ there exists a constant $0 < K_3 < \infty$ such that

$$\|f(x,i)\|^2 \vee \|g(x,i)\|^2 \leq K_3(1 + \|x\|^2). \tag{13}$$

Under Assumptions 2 and 3 there exists a unique solution to (1) on each $[0, T]$: See Mao and Yuan (2006, Theorem 3.16).

It is possible to replace the linear growth condition with a weaker constraint. If Assumption 2 holds, so that f and g satisfy a local Lipschitz condition, and they also satisfy the following monotonicity condition

Assumption 4 There exists a positive constant K such that for all $(x, i) \in \mathbb{R}^d \times [0, T] \times \mathbb{S}$,

$$x^T f(x,i) + \frac{1}{2}\|g(x,i)\|^2 \leq K(1 + \|x\|^2). \tag{14}$$

Then there exists a unique solution X to Eq. (1) on each $[0, T]$. Under the same conditions (Assumption 2 and either Assumption 3 or 4) solutions of (1) also satisfy uniform moment bounds and Hölder regularity bounds: for details see Mao and Yuan (2006).

In our main strong convergence proof we will make use of the following continuous form of the Gronwall inequality (see, for example Mao 2007, Theorem 8.1):

Lemma 1 *Let $T > 0$ and $c \geq 0$. Let u be a Borel measurable bounded nonnegative function on $[0, T]$, and let v be a nonnegative integrable function on $0, T$. If*

$$u(t) \leq c + \int_0^t v(s)u(s)ds \quad \textit{for all} \quad 0 \leq t \leq T,$$

then

$$u(t) \leq c \exp\left(\int_0^t v(s)ds\right) \quad \textit{for all} \quad 0 \leq t \leq T. \tag{15}$$

3 Adaptive Timestepping for an SDEwMS

In this section we define our adaptive mesh as a variant of the construction in Kelly et al. (2025), and characterise the Euler-Maruyama and Milstein methods method used to generate the approximate solutions for the class of SDEsWMS.

3.1 Structure of an Adaptive Mesh

Let $\{h_{n+1}\}_{n\in\mathbb{N}}$ be a sequence of strictly positive random timesteps with corresponding random times $\{t_n := \sum_{i=1}^n h_i\}_{n\in\mathbb{N}\setminus\{0\}}$, where $t_0 = 0$. We set up a framework by which nodes in this mesh can be generated procedurally as a function of the solution at the start of each step, while also ensuring that the timepoints at which the Markov chain switches between distinct states are included.

Assumption 5 For each term of the sequence of random timesteps $\{h_{n+1}\}_{n\in\mathbb{N}}$ at least one of the following three possibilities holds:

1. there are constant values $h_{\max} > h_{\min} > 0$, $\rho > 1$ such that $h_{\max} = \rho h_{\min}$, and
$$0 < h_{\min} \leq h_{n+1} \leq h_{\max} \leq 1;$$
2. there exists a switching time τ_k, as defined in Sect. 2, such that $t_{n+1} = \tau_k$, so that
$$0 < h_{n+1} = \tau_k - t_n \leq h_{\max} \leq 1;$$
3. $t_{n+1} = T$, so that
$$0 < h_{n+1} = T - t_n \leq h_{\max} \leq 1.$$

In addition we assume each h_{n+1} is $\mathcal{F}_{t_n}$-measurable.

It was shown in Kelly et al. (2025) that as long as h_1 can be computed from the initial data, and successive adaptive timesteps h_{k+1} are computed on each trajectory as a deterministic function of the numerical approximation at time t_k, then each t_n on a mesh satisfying the conditions of Assumption 5 is an $\mathcal{F}_t$-stopping time.

Definition 1 Let $N^{(t)}$ be a random integer such that

$$N^{(t)} := \max\{n \in \mathbb{N}\ \{0\} : t_{n-1} < t\}, \tag{16}$$

and let $N = N^{(t)}$ and $t_N = T$, so that T is always the last point on the mesh.

Let Definition 1 be satisfied with $N^{(t)}_{\text{min}} := \lfloor t/h_{\text{max}} \rfloor$ and $N^{(t)}_{\text{max}} := \lceil t/h_{\text{min}} + \bar{N}_t \rceil$ We ensure that we reach the final time by taking $h_N = T - t_{N-1}$ as our final step, and use the backstop method if $h_N < h_{\text{min}}$. Recall from Sect. 1 that $\bar{N}_t < \infty$ a.s for all $t \in [0, T]$, where $\mathbb{S}$ is a finite set, therefore $N^{(t)}_{\text{max}}$ is an a.s finite random variable.

3.2 *An Explicit Discretisation Scheme for an SDEwMS*

While the main result in this article applies for any strongly convergent numerical method, in this section we give the example of the explicit Euler-Maruyama and Milstein methods, and show how they may be adapted to numerically solve an SDEwMS.

3.2.1 The Explicit Euler-Maruyama Scheme

We may extend (6), presented in Sect. 1, to solve an SDEwMS over an adaptive mesh $\{t_n\}$ as

$$X_{n+1} = X_n + h_{n+1} f(X_n, r(t_n)) + g(X_n, r(t_n))\Delta W_{n+1}. \tag{17}$$

Then the continuous approximation for all $s \in [t_n, t_{n+1}]$ is

$$\bar{X}(s) = X_n + \int_{t_n}^{s} f(\bar{X}(t_n), r(t_n))dr + \int_{t_n}^{s} g(\bar{X}(t_n), r(t_n))dW(r). \tag{18}$$

Notice that since switching times are included in the adaptive mesh as constructed in Sect. 3.1 we can simply fix the value of the Markov chain r to the value observed at the beginning of each step.

3.2.2 The Milstein Scheme

In the case where $d = m = 1$, a continuous form of the Milstein scheme for the SDE (4) may be given as

$$X_{n+1} = X_n + h_{n+1} f(X_n) + g(X_n)\Delta W_{n+1} + \frac{1}{2} g'(X_n) g(X_n) \left(\Delta W_{n+1}^2 - h_{n+1}\right), \tag{19}$$

with corresponding continuous version

$$\begin{aligned}\bar{X}(s) = X_n &+ \int_{t_n}^{s} f(\bar{X}(t_n))dr \\ &+ \int_{t_n}^{s} g(\bar{X}(t_n))dW(r) + \int_{t_n}^{s}\int_{t_n}^{r} g'(\bar{X}(t_n))g(\bar{X}(t_n))dW(p)dW(r).\end{aligned} \tag{20}$$

Again, we extend (19) to solve an SDEwMS over an adaptive mesh $\{t_n\}$ as

$$\begin{aligned}X_{n+1} = X_n &+ h_{n+1} f(X_n, r(t_n)) + g(X_n, r(t_n))\Delta W_{n+1} \\ &+ \frac{1}{2} g'(X_n, r(t_n)) g(X_n, r(t_n)) \left(\Delta W_{n+1}^2 - h_{n+1}\right),\end{aligned} \tag{21}$$

with corresponding continuous version

$$\begin{aligned}\bar{X}(s) = X_n &+ \int_{t_n}^{s} f(\bar{X}(t_n), r(t_n))dr + \int_{t_n}^{s} g(\bar{X}(t_n), r(t_n))dW(r) \\ &+ \int_{t_n}^{s}\int_{t_n}^{r} g'(\bar{X}(t_n), r(t_n))g(\bar{X}(t_n), r(t_n))dW(p)dW(r).\end{aligned} \tag{22}$$

A full description of the Milstein scheme applied to (4) in the more general d-dimensional case with m independent Brownian noise terms requires the simulation of Lévy areas. Details are in Kelly et al. (2023). When $m = d = 1$, the Milstein method (20) may be expressed as a map $\psi : \mathbb{R} \times \mathbb{S} \times [0, T] \times [0, h_{\max}] \to \mathbb{R}$ over each step $[t_n, t_{n+1}]$:

$$\begin{aligned}\psi(x, i, t_n, s - t_n) := x &+ (s - t_n) f(x, i) \\ &+ \int_{t_n}^{s} g(x, i)dW(v) + \int_{t_n}^{s}\int_{t_n}^{v} g'(x, i)g(x, i)dW(p)dW(v).\end{aligned} \tag{23}$$

Finally, note that if we modify the second term on the RHS of (19) to be $h_{n+1} f(X_{n+1})$ we arrive at the implicit Milstein scheme, which will be used later in Sect. 5. A strong convergence analysis of this scheme for SDEs with non-globally Lipschitz continuous coefficients is contained in Higham et al. (2013).

3.3 A Hybrid Adaptive Numerical Scheme: General Form

Constructing our scheme we will use two maps $\mathcal{M} : \mathbb{R}^d \times \mathbb{S} \times [0, T] \times [0, h_{\max}] \to \mathbb{R}^d$ and $\mathcal{B} : \mathbb{R}^d \times \mathbb{S} \times [0, T] \times [0, h_{\max}] \to \mathbb{R}^d$ where $\mathcal{M}$ is the main or default map and $\mathcal{B}$ the backstop map. $\mathcal{M}$ corresponds to an efficient scheme with is convergent as an adaptive method and $\mathcal{B}$ corresponds to a method which is convergent as a fixed step method but which may be inefficient or may induce distortions in solution dynamics, with the aim that $\mathcal{B}$ will be rarely used.

We characterise $\mathcal{M}$ and $\mathcal{B}$ through their action over a single step, requiring them to satisfy a certain mean-square consistency bound.

Definition 2 Let $\{h_{n+1}\}_{n\in\mathbb{N}}$ satisfy Assumption 5. We define the continuous form of hybrid adaptive numerical scheme associated with the timestepping strategy $\{h_{n+1}\}_{n\in\mathbb{N}}$ to be

$$Y(s) = \mathcal{M}(Y(t_n), r(t_n), t_n, s - t_n) \cdot \mathcal{I}_{\{h_{\min} < h_{n+1} \le h_{\max}\}} + \mathcal{B}(Y(t_n), r(t_n), t_n, s - t_n) \cdot \mathcal{I}_{\{h_{n+1} \le h_{\min}\}}. \tag{24}$$

Now define the global error of the scheme.

Definition 3 For any $s \in [0, T]$, we define the global error acquired by the scheme (24) to be

$$E(s) := X(s) - Y(s) \tag{25}$$

where $X(s)$ satisfies (1) and $Y(s)$ satisfies (24).

Then we make the following assumption

Assumption 6 For $s \in [t_n, t_{n+1}]$, $n \in \mathbb{N}$ and any $x, y \in \mathbb{R}^d$, let

$$\begin{aligned} E_{\mathcal{M}}^{x,y,i}(s) &:= x - \mathcal{M}(y, i, t_n, s - t_n); \\ E_{\mathcal{B}}^{x,y,i}(s) &:= x - \mathcal{B}(y, i, t_n, s - t_n). \end{aligned} \tag{26}$$

Assume that $E_{\mathcal{M}}^{x,y,i}$ and $E_{\mathcal{B}}^{x,y,i}$ satisfy, for some $\delta > 0$, both $\psi = \mathcal{M}$ or $\psi = \mathcal{B}$, any $i \in \mathbb{S}$, and any a.s. finite $\mathcal{F}_{t_n}$-measurable $\mathbb{R}^d$-valued random variables A_n, B_n

$$\mathbb{E}\big[||E_{\psi}^{A_n,B_n,i}(t_{n+1})||^2|\mathcal{F}_{t_n}\big] \le ||A_n - B_n||^2 + \Gamma_1 \int_{t_n}^{t_{n+1}} \mathbb{E}\big[||E_{\psi}^{A_n,B_n,i}(r)||^2|\mathcal{F}_{t_n}\big]dr + \Gamma_{2,n} h_{n+1}^{2\delta+1}. \tag{27}$$

In (27), for any $h_{n+1} \le h_{\min}$ in the case where $\psi = \mathcal{B}$ and for any $h_{\min} < h_{n+1} \le h_{\max}$ in the case where $\psi = \mathcal{M}$, $\Gamma_1 < \infty$ is a constant, $\Gamma_{2,n}$ is a scalar $\mathcal{F}_{t_n}$-measurable random variable with finite expectation denoted Γ_2, and both Γ_1 and $\Gamma_{2,n}$ are independent of $h_{\max}$.

3.4 A Hybrid Adaptive Numerical Scheme: Particular Form

To ensure the appropriate bound to satisfy Assumption 6 we use a particular kind of adaptive time-stepping strategy such that, whenever $h_{\min} < h_{n+1} \leq h_{\max}$,

$$||Y(t_n)|| < R. \tag{28}$$

Lemma 2 *Let $Y(s)$ be the hybrid scheme* (24). *Fix $n = 0, ..., N-1$ and $k > 0$. Then, in the event that $h_{min} < h_{n+1} \leq h_{max}$, h_{n+1} satisfies* (28)

$$h_{n+1} = \left(h_{min} \vee \left(\frac{h_{max}}{||Y(t_n)||^{1/k}} \wedge h_{max}\right)\right) \wedge \left(\tau_{\bar{N}_{t_n}+1} - t_n\right). \tag{29}$$

For a proof, see Lemma 4.2 in Kelly et al. (2025). Since the mesh $\{t_n\}$ has been constructed so that the value of the Markov chain r cannot change between mesh-points, conditions under which the Milstein map (23) satisfies (27) over a single step with $\delta = 1$ are independent of i and the same as those provided in Kelly et al. (2025, Sect. 4).

4 Main Result

For any adaptive numerical scheme composed of main and backstop maps that satisfy the bound (27) in Assumption 6, our main result demonstrates strong L_2 convergence when extended to an SDEwMS via a switching-time adapted mesh. The proof is a modification of that of Kelly et al. (2025, Theorem 4.1), though the change of setting from SJDEs to SDEwMS allows for the analysis to be simplified in places. By using the Milstein method (20) we can construct a specific scheme of order $\delta = 1$.

Theorem 1 *Let $(X(t))_{t\in[0,T]}$ be a solution of* (1) *with initial value $X(0) = x_0 \in \mathbb{R}^d$ and suppose that Assumptions 1, 2, and 4 hold. Let Y be an adaptive numerical scheme as characterised in Definition 2 with $Y_0 = x_0$ such that Assumption 6 holds with some $\delta > 0$. Then there exists a constant $C(T) > 0$ such that*

$$\max_{t\in[0,T]} \left(\mathbb{E}\left[\|X(t) - Y(t)\|^2\right]\right)^{\frac{1}{2}} \leq C(T)h_{\max}^{\delta}, \tag{30}$$

where C does not depend on $h_{\max}$.

Proof The proof naturally divides into four steps which successively show how conditional mean-square errors accumulate, first over the steps between successive switching times, then from the initial time to the final switching time before a fixed time t, then from the last switching time to t, and finally over the full interval $[0, t]$. We may then apply the Gronwall inequality (see Lemma 1) to complete the proof.

Step 1: cumulative error bound over the interval between successive switches. Over a single step, we obtain from (27) in Assumption 6 the following error bound regardless of whether the scheme uses the main map or the backstop map over a single step of length $0 < h_{n+1} \leq h_{\max}$, and regardless of whether or not a change in the state of the Markov chain r occurs at either endpoint of the timestep: a.s,

$$\mathbb{E}\left[\|E(t_{n+1})\|^2|\mathcal{F}_{t_n}\right] \leq \|E(t_n)\|^2 + \Gamma_1 \int_{t_n}^{t_{n+1}} \mathbb{E}\left[\|E(u)\|^2|\mathcal{F}_{t_n}\right] du + \Gamma_{2,n} h_{n+1}^{2\delta+1}. \tag{31}$$

Let τ_k and τ_{k+1} be successive switching times. Subtract the first term on the RHS of (31) from both sides of the inequality and sum both sides over all the steps from τ_k to τ_{k+1}. Then we get

$$\begin{aligned}
\mathcal{K}_k :=& \sum_{n=N^{(\tau_k)}}^{N^{(\tau_{k+1})}-1} \left(\mathbb{E}\left[\|E(t_{n+1}\|^2|\mathcal{F}_{t_n}\right] - \|E(t_n)\|^2\right)\mathcal{I}_{\{N^{(\tau_{k+1})}>n\}} \\
\leq& \sum_{n=N^{(\tau_k)}}^{N^{(\tau_{k+1})}-1} \left(\Gamma_1 \int_{t_n}^{t_{n+1}} \mathbb{E}[\|E(u)\|^2|\mathcal{F}_{t_n}]du + \Gamma_{2,n} h_{n+1}^{2\delta+1}\right)\mathcal{I}_{\{N^{(\tau_{k+1})}>n\}} \\
=&\Gamma_1 \sum_{n=N^{(\tau_k)}}^{N^{(\tau_{k+1})}-1} \int_{t_n}^{t_{n+1}} \mathbb{E}\left[\|E(u)\|^2|\mathcal{F}_{t_n}\right]\mathcal{I}_{\{N^{(\tau_{k+1})}>n\}} du \\
&+ \sum_{n=N^{(\tau_k)}}^{N^{(\tau_{k+1})}-1} \Gamma_{2,n} h_{n+1}^{2\delta+1}\mathcal{I}_{\{N^{(\tau_{k+1})}>n\}}, \quad a.s.
\end{aligned}$$

By Definition 1, the filtration $\mathcal{F}_{t_n}$ may be written as $\mathcal{F}_{t_{N(u)-1}}$, for all $u \in [t_n, t_{n+1})$ and $n = 0, \ldots, N^{(u)} - 1$, so we can rewrite the inequality as

$$\mathcal{K}_k \leq \Gamma_1 \int_{\tau_k}^{\tau_{k+1}} \mathbb{E}\left[\|E(u)\|^2|\mathcal{F}_{t_{N(u)-1}}\right] dr + \sum_{n=N^{(\tau_k)}}^{N^{(\tau_{k+1})}-1} \Gamma_{2,n} h_{n+1}^{2\delta+1}, \quad a.s, \tag{32}$$

where all indicator variables on the RHS have been bounded above by 1.

Step 2: cumulative error from time $t = 0$ to the time of the last switch before time t. Fix any $t \in [0, T]$. Summing over k for all switching times on the interval $(0, t]$ and denoting $Q(t) := \sum_{k=0}^{\bar{N}_t-1} \mathcal{K}_k$, we have a.s,

$$Q(t) \leq \Gamma_1 \sum_{k=0}^{\bar{N}_t-1} \int_{\tau_k}^{\tau_{k+1}} \mathbb{E}\left[\|E(u)\|^2|\mathcal{F}_{t_{N(u)-1}}\right] du + \sum_{k=0}^{\bar{N}_t-1} \sum_{n=N^{(\tau_k)}}^{N^{(\tau_{k+1})}-1} \Gamma_{2,n} h_{n+1}^{2\delta+1}. \tag{33}$$

Set $\tau_0 = 0$ to include time from 0 to the first switch τ_1 and take the expectation conditional upon $\mathcal{H}_T$

$$\mathbb{E}\left[Q(t)|\mathcal{H}_T\right] \leq \Gamma_1 \mathbb{E}\left[\sum_{k=0}^{\bar{N}_t-1} \int_{\tau_k}^{\tau_{k+1}} \mathbb{E}\left[\|E(u)\|^2 | \mathcal{F}_{t_{N(u)-1}}\right] du \middle| \mathcal{H}_T\right] \\ + \mathbb{E}\left[\sum_{k=0}^{\bar{N}_t-1} \sum_{n=N^{(\tau_k)}}^{N^{(\tau_{k+1})}-1} \Gamma_{2,n} h_{n+1} \middle| \mathcal{H}_T\right] h_{\max}^{2\delta}, \quad a.s.$$

Since $\bar{N}_t$ is $\mathcal{H}_T$ measurable we can take all of the summations on the RHS out of the conditional expectations.

$$\mathbb{E}\left[Q(t)|\mathcal{H}_T\right] \leq \Gamma_1 \sum_{k=0}^{\bar{N}_t-1} \mathbb{E}\left[\int_{\tau_k}^{\tau_{k+1}} \mathbb{E}\left[\|E(u)\|^2 | \mathcal{F}_{t_{N(u)-1}}\right] du \middle| \mathcal{H}_T\right] \\ + \sum_{k=0}^{\bar{N}_t-1} \mathbb{E}\left[\sum_{n=N^{(\tau_k)}}^{N^{(\tau_{k+1})}-1} \Gamma_{2,n} h_{n+1} \middle| \mathcal{H}_T\right] h_{\max}^{2\delta}, \quad a.s.$$

Since $\mathcal{H}_T \subseteq \mathcal{F}_{t_{N(u)-1}}$ for any $u \in [0, T]$, an application of the tower property of conditional expectations yields

$$\mathbb{E}\left[Q(t)|\mathcal{H}_T\right] \leq \Gamma_1 \int_0^{\tau_{\bar{N}_t}} \mathbb{E}\left[\|E(u)\|^2 | \mathcal{H}_T\right] du \\ + \sum_{k=0}^{\bar{N}_t-1} \mathbb{E}\left[\sum_{n=N^{(\tau_k)}}^{N^{(\tau_{k+1})}-1} \Gamma_{2,n} h_{n+1} \middle| \mathcal{H}_T\right] h_{\max}^{2\delta}, \quad a.s. \tag{34}$$

Step 3: error bound from the time of the last switch to time t. Fix $t \in [0, T]$. If $t_{N^{(t)}} = t$ this step can be omitted. Suppose that $t_{N^{(t)}} < t$. Replacing t_n with $t_{N^{(t)}-1}$ and t_{n+1} with t in (31), we have a bound on the error over the last step to time t as

$$\mathbb{E}\left[\|E(t)\|^2 | \mathcal{F}_{t_{N(t)-1}}\right] \leq \|E(t_{N^{(t)}-1})\|^2 \\ + \Gamma_1 \int_{t_{N(t)-1}}^{t} \mathbb{E}[\|E(u)\|^2 | \mathcal{F}_{t_{N(t)-1}}] du + \Gamma_{2,n} |t - t_{N^{(t)}-1}|^{2\delta+1}. \tag{35}$$

Now summing up to t with the error bound over the last step satisfying (35) and taking the expectation conditioned on $\mathcal{H}_T$ we have, for all $t \in [0, T]$ and a.s,

$$\mathcal{R}(t) := \mathbb{E}\left[\sum_{n=N^{(\tau_{\tilde{N}_t})}}^{N^{(t)}-2} \left(\mathbb{E}\left[\|E(t_{n+1})\|^2|\mathcal{F}_{t_n}\right] - \|E(t_n)\|^2\right) \mathcal{I}_{\{N^{(t)}>n+1\}}\right.$$
$$\left. + \mathbb{E}\left[\|E(t)\|^2|\mathcal{F}_{t_{N^{(t)}-1}}\right] - \|E(t_{N^{(t)}-1})\|^2 \middle| \mathcal{H}_T\right]$$
$$\leq \Gamma_1 \mathbb{E}\left[\sum_{n=N^{(\tau_{\tilde{N}_t})}}^{N^{(t)}-2} \int_{t_n}^{t_{n+1}} \mathbb{E}\left[\|E(u)\|^2|\mathcal{F}_{t_n}\right] \mathcal{I}_{\{N^{(t)}>n+1\}} du\right. \tag{36}$$
$$\left. + \int_{t_{N^{(t)}-1}}^{t} \mathbb{E}\left[\|E(u)\|^2|\mathcal{F}_{t_{N^{(t)}-1}}\right] du \middle| \mathcal{H}_T\right]$$
$$+ h_{\max}^{2\delta} \mathbb{E}\left[\sum_{n=N^{(\tau_{\tilde{N}_t})}}^{N^{(t)}-2} \Gamma_{2,n} h_{n+1} + \Gamma_{2,N^{(t)}-1} |t - t_{N^{(t)}-1}| \middle| \mathcal{H}_T\right]. \tag{37}$$

Summing the integrals on the RHS of (37), bounding the indicators on the RHS above by 1, and again applying the tower property of conditional expectations we get

$$\mathcal{R}(t) \leq \Gamma_1 \int_{\tau_{\tilde{N}_t}}^{t} \mathbb{E}\left[\|E(u)\|^2|\mathcal{H}_T\right] du$$
$$+ h_{\max}^{2\delta} \mathbb{E}\left[\sum_{n=N^{(\tau_{\tilde{N}_t})}}^{N^{(t)}-2} \Gamma_{2,n} h_{n+1} + \Gamma_{2,N^{(t)}-1} |t - t_{N^{(t)}-1}| \middle| \mathcal{H}_T\right], \; t \in [0, T], \; a.s. \tag{38}$$

Step 4: cumulative error bound from time $t = 0$ to time t. Adding the error bound associated with the steps that occur after the last switch (38) to the error bound at the last switch (34) we have a.s,

$$\mathbb{E}\left[Q(t)|\mathcal{H}_T\right] + \mathcal{R}(t) \leq \Gamma_1 \int_0^t \mathbb{E}\left[\|E(u)\|^2|\mathcal{H}_T\right] du$$
$$+ h_{\max}^{2\delta} \mathbb{E}\left[\sum_{n=0}^{N^{(t)}-2} \Gamma_{2,n} h_{n+1} + \Gamma_{2,N^{(t)}-1} |t - t_{N^{(t)}-1}| \middle| \mathcal{H}_T\right], \quad t \in [0, T]. \tag{39}$$

Next lets simplify the LHS of (39):

$$
\begin{aligned}
&\mathbb{E}\left[Q(t)\middle|\mathcal{H}_T\right]+\mathcal{R}(t)\\
&=\mathbb{E}\left[\sum_{k=0}^{\bar{N}_t-1}\left(\sum_{n=N^{(\tau_k)}}^{N^{(\tau_k)}-1}\left(\mathbb{E}\left[\|E(t_{n+1})\|^2\middle|\mathcal{F}_{t_n}\right]-\|E(t_n)\|^2\right)\mathcal{I}_{\{N^{(\tau_{k+1})}>n\}}\right.\right.\\
&\qquad\left.+\sum_{n=N^{(\tau_{\bar{N}_t})}}^{N^{(t)}-2}\left(\mathbb{E}\left[\|E(t_{n+1})\|^2\middle|\mathcal{F}_{t_n}\right]-\|E(t_n)\|^2\right)\mathcal{I}_{\{N^{(t)}>n+1\}}\right)\\
&\qquad\left.+\mathbb{E}\left[\|E(t)\|^2\middle|\mathcal{F}_{t_{N^{(t)}-1}}\right]-\|E(t_{N^{(t)}-1})\|^2\middle|\mathcal{H}_T\right]\\
&=\mathbb{E}\left[\sum_{n=0}^{N^{(t)}-2}\left(\mathbb{E}\left[\|E(t_{n+1})\|^2\middle|\mathcal{F}_{t_n}\right]-\|E(t_n)\|^2\right)\mathcal{I}_{\{N^{(t)}>n+1\}}\right.\\
&\qquad\left.+\mathbb{E}\left[\|E(t)\|^2\middle|\mathcal{F}_{t_{N^{(t)}}-1}\right]-\|E(t_{N^{(t)}-1})\|^2\middle|\mathcal{H}_T\right],\ t\in[0,T],\ a.s.
\end{aligned}
$$

As $N^{(t)}_{\max}=\lceil t/h_{\min}+\bar{N}_t\rceil$ is $\mathcal{H}_T$ measurable, we can bound $N^{(t)}$ by $N^{(t)}_{\max}$ and move the sum out of the conditional expectation. Again applying the tower property and now the telescoping sum with $E(0)=0$ we see that

$$
\begin{aligned}
&\mathbb{E}\left[Q(t)|\mathcal{H}_T\right]+\mathcal{R}(t)\\
&=\sum_{n=0}^{N^{(t)}_{\max}-2}\mathbb{E}\left[\mathbb{E}\left[\|E(t_{n+1})\|^2\mathcal{I}_{\{N^{(t)}>n+1\}}|\mathcal{F}_{t_n}\right]-\|E(t_n)\|^2\mathcal{I}_{\{N^{(t)}>n+1\}}\right.\\
&\qquad\left.+\mathbb{E}\left[\|E(t)\|^2\middle|\mathcal{F}_{t_{N^{(t)}}-1}\right]-\|E(t_{N^{(t)}-1})\|^2\middle|\mathcal{H}_t\right]\\
&=\mathbb{E}\left[\|E(t)\|^2|\mathcal{H}_T\right],\ t\in[0,T],\ a.s.
\end{aligned}
$$

Substituting (40) into (39) we have

$$
\begin{aligned}
\mathbb{E}\left[\|E(t)\|^2|\mathcal{H}_T\right]&\le\Gamma_1\int_0^t\mathbb{E}\left[\|E(u)\|^2|\mathcal{H}_T\right]du\\
&+h_{\max}^{2\delta}\mathbb{E}\left[\sum_{n=0}^{N^{(t)}-2}\Gamma_{2,n}h_{n+1}+\Gamma_{2,N^{(t)}-1}|t-t_{N^{(t)}-1}|\middle|\mathcal{H}_T\right],\ t\in[0,T],\ a.s. \qquad (40)
\end{aligned}
$$

Now let's consider the second term on the RHS of (40). On each trajectory $\omega \in \Omega$ we can define a step function $\bar{\Gamma}_2(s)(\omega)$, $s \in [t_n(\omega), t_{n+1}(\omega))$, $n = 0, \cdots, N^{(t)}(\omega)$. Then write the summation as an integral and take the expectation to get

$$\mathbb{E}\left[h_{\max}^{2\delta} \mathbb{E}\left[\sum_{n=0}^{N^{(t)}-2} \Gamma_{2,n} h_{n+1} + \Gamma_{2,N^{(t)}-1} \left| t - t_{N^{(t)}-1} \right| \middle| \mathcal{H}_T \right] \right]$$
$$= h_{\max}^{2\delta} \mathbb{E}\left[\mathbb{E}\left[\int_0^t \bar{\Gamma}_2(s) ds \middle| \mathcal{H}_T \right] \right].$$

Since $\mathbb{E}[\Gamma_{2,n}] < \Gamma_2 < \infty$ and by construction of the step function we see that

$$h_{\max}^{2\delta} \mathbb{E}\left[\mathbb{E}\left[\int_0^t \bar{\Gamma}_2(s) ds \middle| \mathcal{H}_T \right] \right] = h_{\max}^{2\delta} \left[\int_0^t \mathbb{E}\left[\bar{\Gamma}_2(s) \right] ds \right] \le h_{\max}^{2\delta} t \Gamma_2. \tag{41}$$

Substitute (41) back into (40) and take expectations on both sides to get

$$\mathbb{E}\left[\mathbb{E}\left[\|E(t)\|^2 \middle| \mathcal{H}_T \right] \right] \le \mathbb{E}\left[\Gamma_1 \int_0^t \mathbb{E}\left[\|E(u)\|^2 \middle| \mathcal{H}_T \right] du \right] + h_{\max}^{2\delta} t \Gamma_2, \tag{42}$$

and again it follows from the tower property that

$$\mathbb{E}\left[\|E(t)\|^2 \right] \le \Gamma_1 \int_0^t \mathbb{E}\left[\|E(u)\|^2 \right] du + h_{\max}^{2\delta} t \Gamma_2, \quad t \in [0, T]. \tag{43}$$

Now we can apply the Gronwall inequality (see Lemma 1) to get

$$\begin{aligned} \left(\mathbb{E}\left[\|E(t)\|^2 \right] \right)^{\frac{1}{2}} &\le \left(h_{\max}^{2\delta} t \Gamma_2 \exp \Gamma_1 \int_0^t dr \right)^{\frac{1}{2}} \\ &\le \sqrt{\Gamma_2 t \exp \Gamma_1 t} h_{\max}^{\delta} \\ &\le C(t) h_{\max}^{\delta}, \quad t \in [0, T]. \end{aligned} \tag{44}$$

Taking the maximum over $t \in [0, T]$ on both sides, the bound (30) and therefore the statement of the theorem follows.

Remark 1 The error constant C in the bound (30) may depend on the stepsize ratio ρ indirectly through Γ_1 and Γ_2. However C has no dependence on the switching rates λ_i associated with the Markov chain. This is by contrast with the jump case investigated in Kelly et al. (2025), where the error constant depends directly on a Poisson jump intensity rate.

5 Numerical Results

5.1 *Implementation of an Adaptive Numerical Method for SDEwMS*

First, we simulate the trajectory of a given Markov chain r with state space $\mathbb{S}$ and generator matrix Γ (see Eq. (9)). For each sampled $\omega \in \Omega$, we wish to compute the value of $r(\tau_i, \omega)$, along with the respective switching times τ_i, for $i = 0, 1, ..., \bar{N}_T$. For a trajectory that has just transitioned to state S_i, the holding time before transitioning to a different state is an exponentially distributed random variable with parameter λ_i, where $\lambda_i = \gamma_{i1} + \gamma_{i2} + ... + \gamma_{iL} - \gamma_{ii}$. The conditional transition probabilities are computed using the entries of Γ as

$$p_{ij} = \frac{\gamma_{ij}}{\gamma_{i1} + \gamma_{i2} + ... + \gamma_{iL} - \gamma_{ii}}.$$

Then for each state i we can create a probability mass function with these transition probabilities and sample from it to simulate which state S_j is chosen at each switching time. We store the switching times $(\tau_i)_{i\geq 0}$ and the states selected $(S_i)_{i\geq 0}$ as two separate vectors for each trajectory ω to use in the main algorithm.

Then we can begin to construct the mesh, with specific timestepping strategy as described in (29). The step size, h_n, taken for each step dictates whether we implement the main map $\mathcal{M}$ or the backstop map $\mathcal{B}$ as per (24). In this section we choose as our main map the Milstein map (23) and as our backstop map the map associated with the implicit Milstein method. Therefore our adaptive method will be strongly convergent in mean-square with order $\gamma = 1$.

Running this method for a specified number of trajectories ω with initial starting point L_0, we plot the final values on each trajectory $L_T(\omega)$ as a density histogram.

5.2 *Example: A Stochastic Model of Telomere Length with Markovian Switching*

Consider the following variant of the SDE (2)

$$dL(t) = -\left(C(t) + A(t)L(t)^2\right)dt + \sqrt{\frac{1}{3}A(t)L(t)^3}dW(t), \tag{45}$$

where the random pair $(C(t), A(t))_{t\geq 0}$ evolves according to a Markov chain that is independent of W and with state space $\mathbb{S} = \{1, 2, 3, 4\}$. Here, each $i \in \mathbb{S}$ corresponds to an element of the set $\mathbb{S}' = \{(c_1, a_1), (c_1, a_2), (c_2, a_1), (c_2, a_2)\}$, where a_i and c_j, $i, j = 1, 2$ are constants calculated from data in Grasman et al. (2011). There, different estimation methods provided different values for these constants and so we can consider the use of Markovian switching in this context to be a form of robustness testing. The estimates we will use for our numerical analysis are $(c_1, a_1) = (4.5, 0.22 \times 10^{-6})$, and $(c_2, a_2) = (7.5, 0.41 \times 10^{-6})$, and we will switch between the corresponding states in $\mathbb{S}'$ according to r generated with the algorithm described in Sect. 5.1. For the purposes of our numerical demonstration the generator is chosen to be

$$\Gamma = \begin{bmatrix} -0.3 & 0.1 & 0.1 & 0.1 \\ 0.1 & -0.3 & 0.1 & 0.1 \\ 0.1 & 0.1 & -0.3 & 0.1 \\ 0.1 & 0.1 & 0.1 & -0.3 \end{bmatrix}$$

For the adaptive mesh, we set $h_{max} = 3 \times 10^{-2}$, $k = 10$ and $\rho = 15$ throughout.

5.3 Numerical Results: Fixed Initial Length L_0

For our fixed initial length we have chosen $L_0 = 1000$ and $T = 30$. In Fig. 1 we display a histogram of 1000 values, with $\mathbb{E}[L_{30}] \approx 814.33$ base pairs (bp).

In Fig. 2 we display density histograms produced when the parameters are fixed at $(c, a) = (c_1, a_1)$ (top) and $(c, a) = (c_2, a_2)$ (bottom). The means are computed to be $\mathbb{E}[L_{30}] \approx 862.69$ bp and $\mathbb{E}[L_{30}] \approx 770.75$ bp respectively.

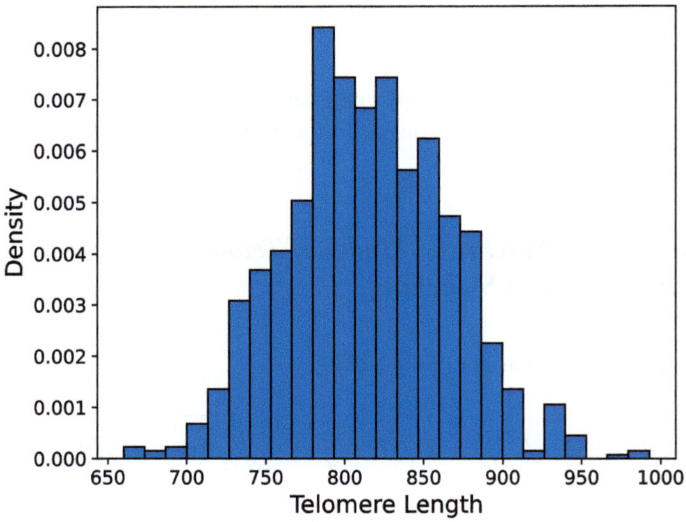

Fig. 1 Histogram of $L_{30}(\omega)$ for 1000 samples

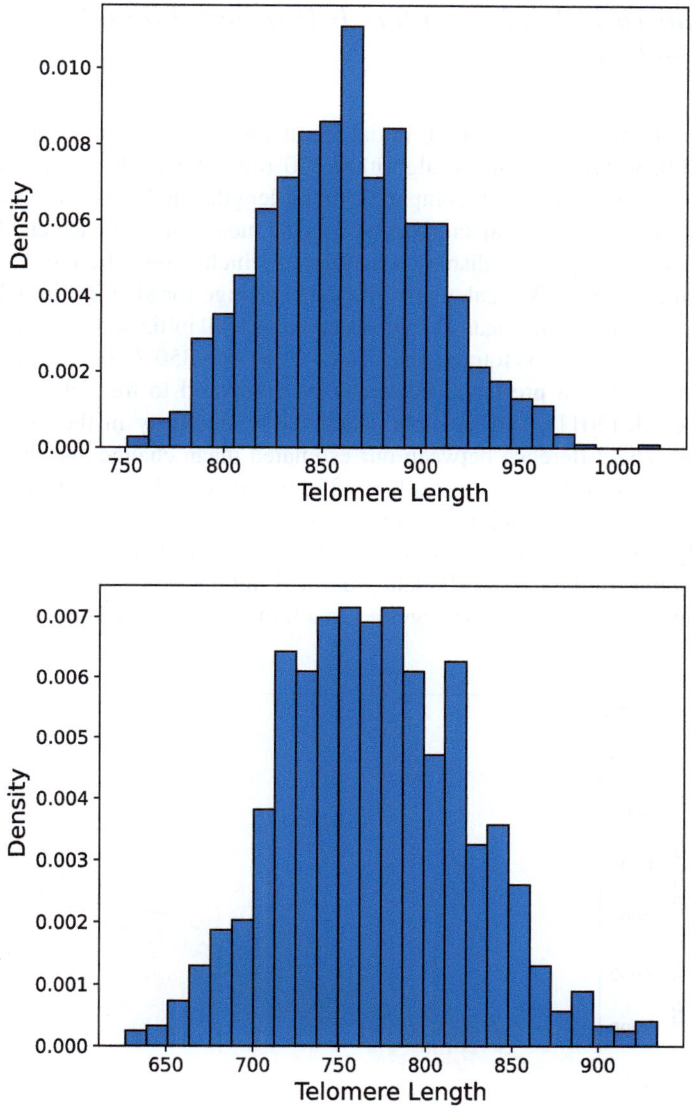

Fig. 2 Histograms of $L_{30}(\omega)$ for 1000 samples for the SDE without switching and choosing $(c, a) = (c_1, a_1)$ (top) and $(c, a) = (c_2, a_2)$ (bottom)

5.4 *Numerical Results: Uniformly Distributed Initial Lengths L_0*

Next we randomly generated 1000 initial lengths from a uniform distribution with range [4000, 8000] and ran the algorithm 100 times for each starting length. In Fig. 3 we display a line graph comparing initial lengths, and for each initial length the final value of a single trajectory as well as the mean value of each end length.

And finally in Fig. 4 we display a histogram which shows the mean change in length of the telomere. We calculated the mean change for all 1000 initial lengths, over 100 simulations. The mean change observed over all initial lengths and across all simulated trajectories was found to be $\mathbb{E}[L_5 - L_{30}] \approx -350.74$ bp. The distribution of the simulated data presented here may be compared to the data presented in Grasman et al. (2011), and we note a qualitative similarity in the mean change histograms. The difference between our estimated mean change and that given in Grasman et al. (2011) ($\mathbb{E}[L_5 - L_{30}] \approx -223.68$ bp) may be the result of using a model with Markovian parameters (including potentially the choice of Γ), as well as possibly reflecting sampling variation and deviations from a uniform distribution of initial values in the real-world data presented in Grasman et al. (2011). A more detailed investigation of this is suggested as a future line of research.

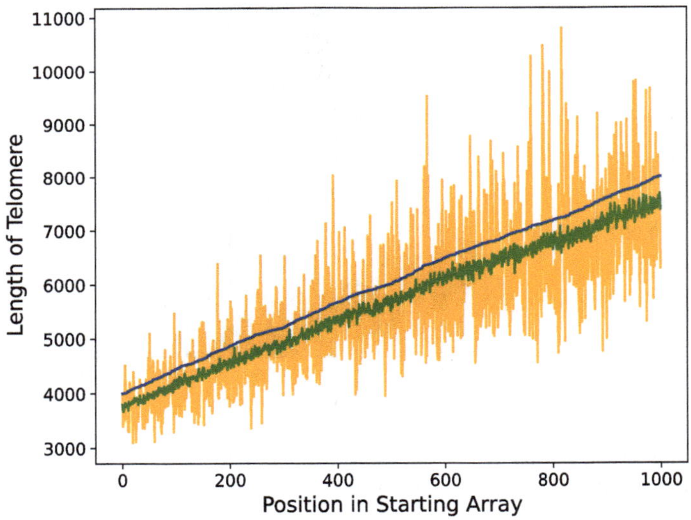

Fig. 3 Initial telomere lengths ordered (blue), actual final length for a single trajectory (orange), and the mean change in length over all trajectories for a given initial value (green)

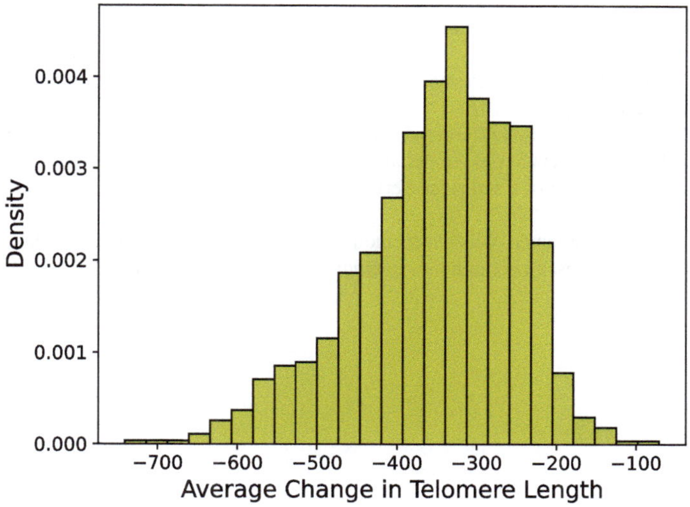

Fig. 4 Density histogram of the mean change in telomere lengths between day 5 and day 30

Acknowledgements CK was supported by the RSE Saltire Research Facilitation Network on Stochastic Differential Equations: Theory, Numerics and Applications (RSE1832).

References

Beyn, W.-J., Isaak, E., Kruse, R.: Stochastic C-stability and B-consistency of explicit and implicit Milstein-type schemes. J. Sci. Comput. **70**, 1042–1077 (2017)

Grasman, J., Salomons, H., Verhulst, S.: Stochastic modeling of length-dependent telomere shortening in Corvus monedula. J. Theor. Biol. **282**, 1–6 (2011)

Guo, Q., Liu, W., Mao, X., Yue, R.: The truncated Milstein method for stochastic differential equations with commutative noise. J. Comput. Appl. Math. **338**, 298–310 (2018)

Higham, D.J., Mao, X., Szpruch, L.: Convergence, non-negativity and stability of a new Milstein scheme with applications to finance. Discrete Continuous Dyn. Syst. B **18**, 2083–2100 (2013)

Hutzenthaler, M., Jentzen, A., Kloeden, P.E.: Strong and weak divergence in finite time of Euler's method for stochastic differential equations with non-globally Lipschitz continuous coefficients. Proc. R. Soc. Lond. A Math. Phys. Eng. Sci. **467**, 1563–1576 (2011)

Hutzenthaler, M., Jentzen, A., Kloeden, P.E.: Strong convergence of an explicit numerical method for SDEs with nonglobally Lipschitz continuous coefficients. Ann. Appl. Prob. **22**, 1611–1641 (2012)

Kelly, C., Lord, G., Sun, F.: Strong convergence of an adaptive time-stepping Milstein method for SDEs with monotone coefficients. BIT Numer. Math. **63** (2023)

Kelly, C., Lord, G., Sun, F.: Strong convergence of a class of adaptive numerical methods for SDEs with jumps **227**, 461–476 (2025)

Kelly, C., Lord, G.J.: Adaptive time-stepping strategies for nonlinear stochastic systems. IMA J. Numer. Anal. **38**, 1523–1549 (2018)

Kelly, C., Lord, G.J., Sun, F.: Strong convergence of an adaptive time-stepping Milstein method for SDEs with monotone coefficients. BIT Numer. Math. **63** (2023)

Mao, X.: Stochastic Differential Equations and Applications, 2nd edn. Woodhead Publishing, Cambridge (2007)
Mao, X.: The truncated Euler-Maruyama method for stochastic differential equations. J. Comput. Appl. Math. **290**, 370–384 (2015)
Mao, X., Yuan, C.: Stochastic Differential Equations with Markovian Switching. Information and Interdisciplinary Subjects Series, Imperial College Press, G - Reference (2006)
Norris, J.R.: Markov Chains, Cambridge Series in Statistical and Probabilistic Mathematics. Cambridge University Press (1997)
Wang, X., Gan, S.: The tamed Milstein method for commutative stochastic differential equations with non-globally Lipschitz continuous coefficients. J. Differ. Eqn. Appl. **19**, 466–490 (2013)

Solution Space Characterisation of Perturbed Linear Functional and Integro-Differential Volterra Convolution Equations: Cesàro Limits

John A. D. Appleby and Emmet Lawless

Abstract In this article we discuss the requirements needed in order to characterise the solution space of perturbed linear integro-differential Volterra convolution equations. We highlight how in general the pointwise behaviour of perturbation functions does not necessarily propagate through to the solution which the classical literature seems to suggest. To illustrate this general idea we show the Cesàro mean of the solution can converge even in cases when the Cesàro mean of the perturbation function diverges. We provide a characterisation of when such convergence takes place and explicitly identify the limit in terms of the problem data. Furthermore we prove how all results can also be applied to perturbed linear functional differential equations.

Keywords Volterra convolution equations · Asymptotic behaviour · Solution space characterisation · Perturbation · Linear integro-differential equations · Cesaro limit

1 Introduction

Analysis of the asymptotic behaviour of both deterministic and stochastic Volterra type integro-differential equations has attracted a lot of attention in recent decades. For deterministic linear equations of non-convolution type, recent works such as Boykov et al. (2024) investigate stability of the zero solution of systems of perturbed Volterra integro-differential equations while for stochastic equations asymptotic mean square stability is addressed in Appleby (2021). Consider the scalar equation,

J. A. D. Appleby
School of Mathematical Sciences, Dublin City University, Dublin, Ireland
e-mail: john.appleby@dcu.ie

E. Lawless (✉)
Department of Mathematics, University of Michigan, Ann Arbor, MI, USA
e-mail: elawless@umich.edu

E. Braverman and A. Swishchuk (eds.), *Delay and Stochastic Differential Equations*, ICIAM2023 Springer Series 12, https://doi.org/10.1007/978-981-95-7310-3_4

$$\dot{x}(t) = \int_{[0,t]} \nu(ds)x(t-s) + f(t), \quad t \geq 0, \tag{1}$$

where ν is a finite signed Borel measure and $x(0) = \xi \in \mathbb{R}$ is a given initial condition. Such convolution equations are in general well understood, to wit, many monographs both classic and modern (such as Brunner 2017; Corduneanu 1973, 1990; Gripenberg et al. 1990) have been produced providing an in dept study to which the reader is referred for more details. The underlying resolvent is crucial in studying (1), it's dynamics are given by,

$$\dot{r}(t) = \int_{[0,t]} \nu(ds)r(t-s), \quad t > 0; \quad r(0) = 1. \tag{2}$$

As shown in Gripenberg et al. (1990, Theorem 3.9) it is necessary to assume $r \in L^1(\mathbb{R}_+; \mathbb{R})$ in order to obtain strong admissibility results, thus in this article we also employ such an assumption.[1] If we consider the unperturbed version of (1), i.e. with $f = 0$, the solution is simply $x(t) = r(t)\xi$, hence the assumption that $r \in L^1(\mathbb{R}_+; \mathbb{R})$ endows the solution with many desirable properties. For example solutions will vanish at infinity, be bounded, $p-$integrable ($p \geq 1$), absolutely continuous etc., a larger list of properties the solution will possess is collected in Gripenberg et al. (1990, Theorem 3.9). It is therefore very natural to ask, can we characterise when such properties are *preserved* after a perturbation is introduced, or in other words can we give necessary and sufficient conditions on the forcing function f to ensure the solution of equation (1) is an element of a particular function space V. Surprisingly the classical theory does not provide an answer to this question, but a solution was proposed by the authors in Appleby and Lawless (2023b). For a given function space V the key for obtaining such a result is a decomposition lemma[2] which states (for some locally integrable function f) if $\int_{\cdot}^{\cdot+\theta} f(s)ds \in V$ for each $\theta \in (0, 1]$ then $f = f_1 + f_2$ where $f_1 \in V$ and $\int_0^{\cdot} f_2(s)ds \in V$. Such a result was proven for $V = BC_0(\mathbb{R}_+; \mathbb{R})$ (space of bounded continuous functions that vanish at infinity) in Gripenberg et al. (1990, Lemma 15.9.2) and by the authors for $V = L^p(\mathbb{R}_+; \mathbb{R})$ for $p \geq 1$ in Appleby and Lawless (2023b, Lemma 2). In this article this takes the form of Lemma 3 which deals with the case that $V = \text{Ces}(\mathbb{R}_+; \mathbb{R})$ (see Definition 1).

It is the authors opinion that the results provided in Appleby and Lawless (2023b) (and also in this work) are by no means restricted to the particular spaces listed above and that the framework developed can be extended to a wide variety of function spaces including, but not restricted to, those spaces found in Gripenberg et al. (1990, Theorem 3.9). It is in this regard that this article should be understood as further evidence that the methods developed in Appleby and Lawless (2023b) can be used to characterise non-standard behaviour in the solution of equation (1).

---

[1] Recall this is equivalent to $z - \hat{\nu}(z) \neq 0$ for $\text{Re}(z) \geq 0$ where $z \in \mathbb{C}$ and $\hat{\nu}$ denotes the Laplace transform of the measure ν Gripenberg et al. (1990, Theorem 3.3.5).

[2] We also require the operation of convolution with L^1 functions to map V into itself and for $r \in V$, but these conditions are known from the classical theory.

Developing a general approach to characterise the behaviour of solutions to (1) is of interest as such results are fundamental when considering more complicated stochastic extensions such as

$$dX(t) = \left(f(t) + \int_{[0,t]} X(t-s)\nu(ds) \right) dt + \sigma(t)dB(t), \tag{3}$$

where B is a one dimensional Brownian motion and $\sigma \in C(\mathbb{R}_+;\mathbb{R})$. The general theory of stochastic Volterra equations has a long history, we mention the following influential articles (Berger and Mizel 1980a; Pardoux and Protter 1990; Protter 1985) for general existence and uniqueness results. Such general equations have also seen prominent use in applications, asymptotic stability of stochastic hereditary systems is considered in Mizel and Trutzer (1984) while Berger and Mizel (1980b) provides numerous applications in physics such as studying the effects of random heat sources on electrical currents. One can make Eq. (3) explicit by fixing a particular measure ν. For example we could consider a problem where the memory decays exponentially and has an instantaneous point mass, i.e. for some Borel set A we have $\nu(A) = \int_A e^{-\lambda u}\,du + a\delta_0(A)$ where δ_0 is a point mass measure at zero, $\lambda > 0$ and $a \in \mathbb{R}\backslash\{0\}$. This leads to the explicit equation

$$dX(t) = \left(f(t) + aX(t) + \int_0^t e^{-\lambda(t-s)}X(s)\,ds \right) dt + \sigma(t)dB(t).$$

For background material on stochastic differential equations see Gikhman and Skorokhod (1972). Equations such as (3), as well as their finite memory counterpart[3] have been subjected to intense study (Appleby 2002, 2004; Appleby and Freeman 2003; Appleby and Riedle 2006; Chan and Williams 1989). Using the framework developed in Appleby and Lawless (2023b), it is possible to give sharp characterisations of the behaviour of sample paths in an *almost sure* sense. This is demonstrated by the authors in Appleby and Lawless (2024) in which results from Appleby and Lawless (2023b) are leveraged to characterise when solutions of (3) are elements of $L^p(\mathbb{R}_+;\mathbb{R})$ for $p \geq 1$ (pathwise) *almost surely* and when the sample paths vanish at infinity *almost surely*. It is expected that results from this article can also be extended in the same manner. The main use of the deterministic results lies in considering the process $Z(t) = X(t) - Y(t)$ where $Y(t)$ is the solution to an SDE, namely: $dY(t) = (f(t) - Y(t))dt + \sigma(t)dB(t)$. With this observation one can show that Z is continuously differentiable and for each fixed path satisfies an equation of type (1) (Appleby and Lawless 2024, Lemma 3). Thus the study of equation (3) reduces to the study of a Markovian SDE and the deterministic equation (1).

The key point outlined in Appleby and Lawless (2023b) and throughout this article is that if one wishes to obtain such characterisation results for solutions to (1), it is not the pointwise behaviour of the function f that matters but rather its integral over compact intervals:

[3] When ν is supported on a compact interval $[0,\tau]$ for some $\tau > 0$.

$$\int_t^{t+\theta} f(s)ds \quad \text{for } \theta \in (0, 1]. \tag{4}$$

To the best of our knowledge consideration of this quantity in the context of the asymptotic behaviour of perturbed dynamical systems, first appeared in papers of Strauss and Yorke (1967a, b). In many situations it is the case that the forcing function may not be well behaved in a pointwise sense but the smoothing effect of integration drastically improves the situation. For instance in Appleby and Lawless (2023a), the authors provide an example of a continuous function, in which $\limsup_{t\to\infty} f(t) = +\infty$, and yet,

$$\lim_{t\to\infty} \int_t^{t+\theta} f(s)ds = 0 \text{ for each } \theta \in (0, 1].$$

It is shown in Appleby and Lawless (2023b) that given such a perturbation function f, solutions of (1) obey $x(t) \to 0$ as $t \to \infty$, despite the fact that the perturbation function has a diverging lim sup. The theory for Volterra integro-differential equations is intricately linked to the theory of Volterra integral equations, however we make the observation that this phenomenon cannot occur with integral equations. By integral equation we refer to the following:

$$x(t) = \int_0^t k(t-s)x(s)ds + f(t), \quad t \geq 0. \tag{5}$$

We highlight that the classical theory for (24) is in fact sharp and in this case it is indeed the pointwise behaviour of the perturbation term that governs rather then the interval average (4). Thus if one considered the integral equation (5) with $k \in L^1(\mathbb{R}_+; \mathbb{R})$ and a perturbation function f as described above, the solution of (5) would have a diverging lim sup while the solution of the integro-differential equation (with the same kernel k) would converge to zero. We also remark that this "*robustness*" of integro-differential equations is also a phenomenon found in so called functional differential equations wherein the system has only a finite memory.

We illustrate this phenomenon once again by selecting a non standard, yet interesting type of convergence. We are concerned with when solutions of (1) and (5) admit a Cesàro type limit. Assuming $r \in L^1(\mathbb{R}_+, \mathbb{R})$, we show

$$\frac{1}{t}\int_0^t \int_s^{s+\theta} f(u)du\, ds \longrightarrow C\theta \text{ as } t \to \infty \quad \text{for each } \theta \in (0, 1], \tag{6}$$

is equivalent to

$$\frac{1}{t}\int_0^t x(s)ds \longrightarrow \frac{-C}{\nu(\mathbb{R}_+)} \in \mathbb{R} \text{ as } t \to \infty, \tag{7}$$

where $C \in \mathbb{R}$ and x is the solution of (1). Additionally we show that if you alter condition (6) by imposing that f itself must posses a Cesàro limit then this imposes necessary restrictions on the Cesàro limit of the derivative of x. We show,

$$\frac{1}{t}\int_0^t f(s)ds \longrightarrow C \in \mathbb{R} \text{ as } t \to \infty \tag{8}$$

is equivalent to

$$\lim_{t\to\infty}\frac{1}{t}\int_0^t x(s)ds = \frac{-C}{\nu(\mathbb{R}_+)}; \quad \lim_{t\to\infty}\frac{1}{t}\int_0^t \dot{x}(s)ds = 0. \tag{9}$$

In the latter case we note the condition on the Cesàro limit of the derivative cannot be dropped. This highlights the usual situation that conditions imposed directly on the forcing function f are in general *sufficient* to obtain information about the solution, but not *necessary*. However this is once again not the case for solutions to the integral equation (5), in fact condition (8) is equivalent to solutions of (5) having a Cesàro limit which in this case is equal to $C(1 - 1/\nu(\mathbb{R}_+))$. To the best of our knowledge there has been no analysis of the time average of solutions to (1) however we mention analogous equations in discrete time have received some attention (Appleby and Patterson 2016, 2017). Despite this apparent lack of literature there has been much work in the context of admissibility theory for so called Marcinkiewicz spaces which are intimately linked to Cesàro limits of functions. For a brief modern introduction to such spaces we recommend the book by Corduneanu Corduneanu (2009, pp.41–48) or for a more comprehensive study (Bertrandias 1966), while for classical admissibility results regarding integral equations we mention (Beneš 1965).

2 Mathematical Preliminaries

Let $p \in [1, \infty)$ and $f : \mathbb{R}_+ \to \mathbb{R}$. We say $f \in L^p(\mathbb{R}_+; \mathbb{R})$ if $\int_{\mathbb{R}_+} |f(s)|^p ds < +\infty$ and $f \in L^p_{loc}(\mathbb{R}_+; \mathbb{R})$ if $\int_K |f(s)|^p ds < +\infty$, where $K \subset \mathbb{R}_+$ is compact and both integrals are understood in the Lebesgue sense. If $M(\mathbb{R}_+; \mathbb{R})$ is the space of finite signed Borel measures on $\mathbb{R}_+$, and $\nu \in M(\mathbb{R}_+; \mathbb{R})$, we consider the halfline Volterra equation given by

$$\dot{x}(t) = \int_{[0,t]} x(t-s)\nu(ds) + f(t), \quad t \geq 0; \quad x(0) = \xi \in \mathbb{R}. \tag{10}$$

We say x is a solution of (10) on an interval $(0, T]$ whenever x is locally absolutely continuous, satisfies the initial condition and obeys the dynamics in (10) for almost all $t \in (0, T]$. We stipulate throughout, and *without further reference*, that $f \in L^1_{loc}(\mathbb{R}_+; \mathbb{R})$, which guarantees a solution x exists (see Gripenberg et al. 1990, Theorem 3.3.3). In particular, the solution satisfies a variation of constants formula

$$x(t) = r(t)\xi + \int_0^t r(t-s)f(s)ds, \quad t \geq 0, \tag{11}$$

where r is the so–called differential resolvent of ν, which is the unique absolutely continuous function from $\mathbb{R}_+$ to $\mathbb{R}$, satisfying

$$\dot{r}(t) = \int_{[0,t]} r(t-s)\nu(ds), \quad t > 0; \quad r(0) = 1. \tag{12}$$

If μ is a finite measure on $[0, \infty)$, and $f : [0, \infty) \to \mathbb{R}$ then their convolution is

$$(f * \mu)(t) := \int_{[0,t]} f(t-s)\mu(ds), \quad t \geq 0.$$

The convolution of two functions is defined analogously. Given a measure $\mu \in M(\mathbb{R}_+; \mathbb{R})$, we denote $|\mu|$ as the set function which takes each Borel set $E \subseteq \mathbb{R}_+$ and assigns the total variation of μ on E. Recall the total variation of a measure μ on a set E is given by $|\mu|(E) := \sup_\pi \sum_{j=1}^n |\mu(E_j)|$ where π denotes the collection of finite partitions of the set E. For more details on these definitions see Gripenberg et al. (1990, Sect. 3.5). With a slight abuse of terminology we shall call $|\mu|$ the total variation measure of μ. The following standard estimate (Gripenberg et al. 1990, Theorem 3.5.6) will be heavily utilised:

$$|(f * \mu)(t)| \leq \int_{[0,t]} |f(t-s)||\mu|(ds).$$

The reader should keep in mind that all results presented for solutions of equation (10) can be emulated for the finite memory problem, which is a type of functional differential equation. In this instance we would fix a constant $\tau > 0$ and consider

$$\dot{x}(t) = \int_{[-\tau,0]} x(t+u)\mu(du) + f(t), \quad t \geq 0; \quad x(t) = \psi(t), \quad t \leq 0 \tag{13}$$

where $\mu \in M([-\tau, 0]; \mathbb{R})$ and $\psi \in C([-\tau, 0]; \mathbb{R})$. With the following convention one can essentially use identical proofs for results regarding equations (10) and (13). For any subset of the real line, write $-E := \{x \in \mathbb{R} : -x \in E\}$. If $\mu \in M([-\tau, 0]; \mathbb{R})$, we can construct a $\tilde{\mu} \in M([0, \infty); \mathbb{R})$ by writing

$$\begin{aligned} &\tilde{\mu}(E) = \mu(-E), \quad \text{for any Borel set } E \subseteq [0, \tau], \\ &\tilde{\mu}(E) = 0, \quad \text{for any Borel set } E \text{ with } E \cap [0, \tau] = \emptyset. \end{aligned} \tag{14}$$

Let $g : \mathbb{R} \to \mathbb{R}$ be such that $g(t) = 0$ for all $t < 0$. With this construction we have

$$\int_{[-\tau,0]} g(t+s)\mu(ds) = \int_{[0,t]} g(t-s)\tilde{\mu}(ds) = (g * \tilde{\mu})(t), \quad t \geq 0.$$

With this notation the resolvent of equation (13) obeys,

$$\dot{r}_\tau(t) = \int_{[0,t]} \tilde{\mu}(ds) r_\tau(t-s), \quad t > 0; \quad r_\tau(0) = 1; \quad r_\tau(t) = 0 \quad t < 0. \tag{15}$$

Here we use the notation r and r_τ to distinguish between the resolvent of the Volterra equation (10) and the functional equation (13) respectively. If we define $F(t) := \int_{[-\tau,0]} \left(\int_s^0 r_\tau(t+s-u)\psi(u)du \right) \mu(ds)$, the solution of (13) has representation

$$x(t, \psi) = r_\tau(t)\psi(0) + F(t) + \int_0^t r_\tau(t-s) f(s) ds, \tag{16}$$

for $t \geq 0$. The assumption $r_\tau \in L^1(\mathbb{R}_+; \mathbb{R})$ ensures that $F = O(e^{-\alpha t})$ for some $\alpha > 0$, this follows from Theorem I.5.4 and Corollary I.5.5 in Diekmann et al. (1995) paired with Lemma 2.1 in Appleby et al. (2009). Thus this extra term which does not appear in the solution for the Volterra equation (11) will not have any contribution towards the Cesàro mean. As we will be primarily concerned with Cesàro limits of functions we provide a precise definition.

Definition 1 We define the space of Cesàro functions as follows:

$$\mathrm{Ces}(\mathbb{R}_+; \mathbb{R}) := \left\{ f \in L^1_{loc}(\mathbb{R}_+; \mathbb{R}) : \lim_{t\to\infty} \frac{1}{t} \int_0^t f(s)ds = C \text{ for some } C \in \mathbb{R} \right\}. \tag{17}$$

Additionally we say f admits a Cesàro limit if $f \in \mathrm{Ces}(\mathbb{R}_+; \mathbb{R})$.

Remark 1 Clearly all integrable functions admit a trivial Cesàro limit. It is in this regard that the extra term (F) appearing in (16) does not provide any additional difficulties. To see this consider $y(t) = x(t, \psi) - F(t)$, then y is the solution of the Volterra integro-differential equation,

$$\dot{y}(t) = \int_{[0,t]} y(t-s)\tilde{\mu}(ds) + f(t), \quad t \geq 0; \quad y(0) = \psi(0).$$

Thus we can apply the theory developed for Eq. (10) and noting as $F(t)$ is integrable, the Cesàro limits of $y(t)$ and $x(t, \psi)$ agree.

As mentioned we shall always employ the assumption that the resolvent is integrable. The exact value of the integral of r and r_τ are needed in many subsequent proofs so we include this as a lemma.

Lemma 1 *Let r and r_τ be the solutions to equations* (12) *and* (15) *respectively and assume $r, r_\tau \in L^1(\mathbb{R}_+;\mathbb{R})$. Then,*

$$\int_0^\infty r(s)ds = \frac{-1}{\nu(\mathbb{R}_+)}; \quad \int_0^\infty \dot{r}(s)ds = -1;$$
$$\int_0^\infty r_\tau(s)ds = \frac{-1}{\mu([-\tau,0])}; \quad \int_0^\infty \dot{r}_\tau(s)ds = -1.$$

Proof of Lemma 1 We only provide the proof for r as if one employs the convention outlined by (14) then the result for r_τ follows immediately. We note the assumption $r \in L^1(\mathbb{R}_+;\mathbb{R})$ guarantees $r \to 0$ as $t \to \infty$. Thus integrating equation (12) over the interval $[0, T]$ we see, $r(T) = 1 + \int_0^T \int_{[0,t]} r(t-s)\nu(ds)dt$. Sending $T \to \infty$ and applying Fubini's theorem we obtain,

$$\begin{aligned} -1 = \int_0^\infty \int_{[0,t]} r(t-s)\nu(ds)dt &= \int_{[0,\infty]} \int_t^\infty r(t-s)dt\, \nu(ds) \\ &= \int_0^\infty r(s)ds\nu(\mathbb{R}_+). \end{aligned}$$

To show the second assertion we integrate (12) over $\mathbb{R}_+$ and apply the above calculation, i.e. $\int_0^\infty \dot{r}(t)dt = \int_0^\infty \int_{[0,t]} r(t-s)\nu(ds)dt = -1$. □

3 Cesàro functions

In this section we discuss some facts about members of the function space $\mathrm{Ces}(\mathbb{R}_+;\mathbb{R})$. The main results of this paper rely on a decomposition of the forcing function f. The general result has the following flavour, let V represent some function space of interest. If $\int_\cdot^{\cdot+\theta} f(s)ds \in V$ for each $\theta \in (0, 1]$ then $f = f_1 + f_2$ such that $f_1 \in V$ and $\int_0^\cdot f_2(s)ds \in V$. Successfully improving the classical admissibility theory is completely determined by whether or not one can prove such a decomposition result. This is accomplished by Lemmas 2 and 3 below for $V = \mathrm{Ces}(\mathbb{R}_+;\mathbb{R})$.

Lemma 2 *Assume $f \in L^1_{loc}(\mathbb{R}_+;\mathbb{R})$. Suppose for all $\theta \in (0, 1]$, $t \mapsto \int_t^{t+\theta} f(s)ds \in \mathrm{Ces}(\mathbb{R}_+;\mathbb{R})$, i.e.*

$$\lim_{t\to\infty} \frac{1}{t} \int_0^t \int_s^{s+\theta} f(u)duds = C(\theta) \in \mathbb{R}.$$

Then for any fixed θ,

$$\sup_{\substack{t\in[1,\infty)\\ T\in[0,\theta]}} \left| \frac{1}{t} \int_0^t \int_s^{s+T} f(u)duds - C(T) \right| < \infty.$$

Proof of Lemma 2 Introduce for some $m \in \mathbb{N}$, the set,

$$Q_m := \left\{T \in [0,\theta]: \left|\frac{1}{t}\int_0^t \int_s^{s+T} f(u)duds - C(T)\right| \le 1, \text{ for all } t \ge m\right\}.$$

First we show this set is measurable. The mapping $t \mapsto \left|\frac{1}{t}\int_0^t \int_s^{s+T} f(u)duds - C(T)\right|$ is continuous thus,

$$\begin{aligned} Q_m &= \left\{T \in [0,\theta]: \left|\frac{1}{t}\int_0^t \int_s^{s+T} f(u)duds - C(T)\right| \le 1, \text{ for all } t \in \mathbb{Q} \cap [m,\infty)\right\} \\ &= \bigcap_{t \in \mathbb{Q}\cap[m,\infty)} \left\{T \in [0,\theta]: \left|\frac{1}{t}\int_0^t \int_s^{s+T} f(u)duds - C(T)\right| \le 1\right\}. \end{aligned}$$

Now for each t introduce, $F_t(T) : [0,\theta] \to \mathbb{R}_+ : T \mapsto \left|\frac{1}{t}\int_0^t \int_s^{s+T} f(u)duds - C(T)\right|$. Then we can write,

$$Q_m = \bigcap_{t \in \mathbb{Q}\cap[m,\infty)} F_t(T)^{-1}([0,1]).$$

Thus to show Q_m is measurable we need only show that $F_t(T)$ is a measurable function, but this is immediate as $C(T)$ is the pointwise limit of measurable functions and thus measurable, and the mapping $T \mapsto \frac{1}{t}\int_0^t \int_s^{s+T} f(u)duds$ is continuous, thus $F_t(T)$ is a measurable function and so Q_m is a well defined measurable set. Additionally it follows from our supposition that $\bigcup_{m=1}^{\infty} Q_m = [0,\theta]$, thus there must exists an $m' \in \mathbb{N}$ such that $Q_{m'}$ has non zero Lebesgue measure. Fix such an m', then by Lemma 15.9.3 in Gripenberg et al. (1990), it must by the case that the set $Q_{m'} - Q_{m'} = \{T - T' : T, T' \in Q_{m'}\}$ contains an open interval $(-\varepsilon, \varepsilon)$. Without loss of generality take $T, T' \in Q_{m'}$ such that $T > T'$, then we have

$$\begin{aligned} &\left|\frac{1}{t}\int_0^t \int_s^{s+T-T'} f(u)duds - C(T-T')\right| \\ &\quad = \left|\frac{1}{t}\int_0^t \int_s^{s+T-T'} f(u)duds - C(T) + C(T')\right| \\ &\quad \le \left|\frac{1}{t}\int_0^t \int_{s-T'}^{s+T-T'} f(u)duds - C(T)\right| + \left|\frac{1}{t}\int_0^t \int_{s-T'}^{s} f(u)duds - C(T')\right| \\ &\quad \le 2 \end{aligned}$$

for all $t \geq m' + \theta$. To see why the first equality holds consider

$$\frac{1}{t}\int_0^t \int_s^{s+T-T'} f(u)duds = \frac{1}{t}\int_0^t \int_{s-T'}^{s+T-T'} f(u)duds - \frac{1}{t}\int_0^t \int_{s-T'}^{s} f(u)duds.$$

Taking limits on both sides yields

$$C(T - T') = C(T) - C(T'). \tag{18}$$

Now as the above bound holds for all $T \in Q_{m'} - Q_{m'}$, we have in particular for all $T \in [0, \varepsilon)$,

$$\left|\frac{1}{t}\int_0^t \int_s^{s+T} f(u)duds - C(T)\right| \leq 2, \quad \text{for all } t \geq m' + \theta.$$

Thus we have

$$\sup_{\substack{t\in[1,\infty)\\ T\in[0,\theta]}} \left|\frac{1}{t}\int_0^t \int_s^{s+T} f(u)duds - C(T)\right| < \infty.$$

□

Remark 2 The identity (18) tells us that when the mapping $t \mapsto \int_t^{t+\theta} f(s)ds \in \text{Ces}(\mathbb{R}_+; \mathbb{R})$, the limit is an additive function of the parameter θ. This along with measurability ensures the limit has the form $C\theta$ for some $C \in \mathbb{R}$, Bingham et al. (1989, Theorem 1.1.8). In the sequel we shall always write the limit in this explicit form.

Lemma 3 *Let $f \in L^1_{loc}(\mathbb{R}_+; \mathbb{R})$. For each $\theta \in (0, 1]$, the following are equivalent,*

(i)

$$\lim_{t\to\infty} \frac{1}{t}\int_0^t \int_s^{s+\theta} f(u)duds = C\theta.$$

(ii) $f = f_1 + f_2$, where $f_1 \in C(\mathbb{R}_+; \mathbb{R})$ and $f_2 \in L^1_{loc}(\mathbb{R}_+; \mathbb{R})$ such that,

$$\lim_{t\to\infty} \frac{1}{t}\int_0^t f_1(u)du = C, \qquad \lim_{t\to\infty} \frac{1}{t}\int_0^t \int_0^s f_2(u)duds = \frac{C\theta}{2}.$$

Proof of Lemma 3 We extend f such that $f(t) = 0$ for $t < 0$. Fix a $\theta \in (0, 1]$ and set $f_1(t) = \frac{1}{\theta}\int_{t-\theta}^t f(u)du$. Thus by hypothesis we immediately obtain the first limit while the local integrability of f ensures f_1 is continuous on the positive real half line. Next we set $f_2 = f - f_1$ and prove the following identity

$$\int_0^t f_2(s)ds = \int_0^t f(s)ds - \int_0^t \frac{1}{\theta}\int_{s-\theta}^s f(u)duds = \frac{1}{\theta}\int_0^\theta \int_{t-v}^t f(s)dsdv. \tag{19}$$

This was proven in the appendix of Appleby and Lawless (2023b) in the case where $\theta = 1$ but for completeness we provide a full proof. We achieve this goal by considering two cases; $t > \theta$ and $t \le \theta$. First let $t > \theta$, integrate f_2 over $[0, t]$ and apply Fubini's theorem to the second integral to obtain,

$$\begin{aligned}\int_0^t f_2(s)ds &= \int_0^t f(s)ds - \int_0^t \frac{1}{\theta}\int_{s-\theta}^s f(u)duds \\ &= \int_0^t f(s)ds - \frac{1}{\theta}\int_{-\theta}^0 \int_0^{u+\theta} f(u)dsdu - \frac{1}{\theta}\int_0^{t-\theta}\int_u^{u+\theta} f(u)dsdu \\ &\quad - \frac{1}{\theta}\int_{t-\theta}^t \int_u^t f(u)dsdu.\end{aligned}$$

Recalling that we extended $f(t)$ to be zero for $t < 0$ shows the second term above is zero. Then evaluating the inner integrals of the last two terms gives,

$$\int_0^t f_2(s)ds = \int_0^t f(s)ds - \int_0^{t-\theta} f(s)ds - \frac{1}{\theta}\int_{t-\theta}^t f(s)(t-u)ds.$$

Now combining all integrals on the right and then applying Fubini's theorem for a final time yields,

$$\int_0^t f_2(s)ds = \frac{1}{\theta}\int_{t-\theta}^t \int_{t-u}^{\theta} f(s)duds = \frac{1}{\theta}\int_0^{\theta}\int_{t-u}^t f(s)dsdu.$$

If $t \in (0, \theta]$ we once again integrate f_2 over $[0, t]$,

$$\begin{aligned}\int_0^t f_2(s)ds &= \int_0^t f(s)ds - \int_0^t \frac{1}{\theta}\int_{s-\theta}^s f(u)duds \\ &= \int_0^t f(s)ds - \int_0^{\theta} \frac{1}{\theta}\int_0^s f(u)duds,\end{aligned}$$

where the second equality follows again from the fact $f(t) = 0$ for $t < 0$. Now we apply Fubini's theorem to the second integral on the right and evaluate the resulting inner integral,

$$\begin{aligned}\int_0^t f_2(s)ds &= \int_0^t f(s)ds - \int_0^{\theta} \frac{1}{\theta}\int_0^s f(u)duds. \\ &= \int_0^t f(s)ds - \frac{1}{\theta}\int_0^t f(u)(t-u)du.\end{aligned}$$

Combining both integrals and applying Fubini's theorem for the final time gives,

$$\int_0^t f_2(s)ds = \frac{1}{\theta}\int_0^t \int_{t-s}^{\theta} f(s)duds = \frac{1}{\theta}\int_0^{\theta}\int_{t-u}^{t} f(s)dsdu.$$

Thus (19) is proven. Now we calculate the Cesàro limit of $\int_0^t f_2(s)ds$.

$$\frac{1}{t}\int_0^t \int_0^s f_2(u)duds = \frac{1}{\theta}\int_0^{\theta} \frac{1}{t}\int_0^t \int_{s-v}^{s} f(u)dudsdv.$$

Subtracting $\frac{C}{\theta}\int_0^{\theta} v\,dv$ from both sides, taking absolute values and applying (19) yields,

$$\left|\frac{1}{t}\int_0^t \int_0^s f_2(u)duds - \frac{C}{\theta}\int_0^{\theta} vdv\right| \le \frac{1}{\theta}\int_0^{\theta}\left|\frac{1}{t}\int_0^t \int_{s-v}^{s} f(u)duds - vC\right| dv.$$

By Lemma 2, the integrand on the RHS is uniformly bounded, thus we can invoke Arzelà's dominated convergence theorem and pass to the limit inside the integral. Hence,

$$\lim_{t\to\infty}\left|\frac{1}{t}\int_0^t \int_0^s f_2(u)duds - \frac{C}{\theta}\int_0^{\theta} v\,dv\right| = 0,$$

as required and for the converse consider,

$$\begin{aligned}\frac{1}{t}\int_0^t \int_{s-\theta}^{s} f(u)duds = \frac{1}{t}\int_0^t \int_{s-\theta}^{s} f_1(u)duds + \frac{1}{t}\int_0^t \int_0^s f_2(u)duds \\ - \frac{1}{t}\int_0^t \int_0^{s-\theta} f_2(u)duds.\end{aligned}$$

The two terms involving f_2 will cancel when we pass to the limit, so we need only focus on the f_1 term. Consider for some $k > 0$ and $t > k$,

$$\begin{aligned}\frac{1}{t}\int_0^t \int_{s-\theta}^{s} f_1(u)duds &= \frac{1}{t}\int_0^t \int_0^{s-\theta} f_1(u)duds - \frac{1}{t}\int_0^t \int_0^s f_1(u)duds \\ &= \frac{1}{t}\int_0^k \int_0^s f_1(u)duds + \frac{1}{t}\int_k^t \int_0^s f_1(u)duds \\ &\quad - \frac{1}{t}\int_0^k \int_0^{s-\theta} f_1(u)duds - \frac{1}{t}\int_k^t \int_0^{s-\theta} f_1(u)duds.\end{aligned}$$

The first and third terms will vanish as $t \to \infty$ so we focus on the second and fourth.

$$\begin{aligned}
\frac{1}{t}\int_k^t \int_0^s f_1(u)duds - \frac{1}{t}\int_k^t \int_0^{s-\theta} f_1(u)duds \\
&= \frac{1}{t}\int_k^t s\left[\frac{\int_0^s f_1(u)du}{s} - C\right]ds + \frac{C}{t}\int_k^t s\,ds \\
&- \frac{1}{t}\int_k^t (s-\theta)\left[\frac{\int_0^{s-\theta} f_1(u)du}{s-\theta} - C\right]ds - \frac{C}{t}\int_k^t (s-\theta)\,ds.
\end{aligned}$$

Combining the second and fourth integrals yields,

$$\theta C\frac{(t-k)}{t} \quad \longrightarrow \quad \theta C \quad \text{as } t \to \infty.$$

Thus to complete the proof we show the terms involving f_1 vanish. We have

$$\frac{1}{t}\int_k^t s\left[\frac{\int_0^s f_1(u)du}{s} - C\right]ds - \frac{1}{t}\int_k^t (s-\theta)\left[\frac{\int_0^{s-\theta} f_1(u)du}{s-\theta} - C\right]ds.$$

Letting $u = s - \theta$ in the second integral and combining the two yields,

$$\frac{1}{t}\int_{t-\theta}^t s\left[\frac{\int_0^s f_1(u)du}{s} - C\right]ds - \frac{1}{t}\int_{k-\theta}^k s\left[\frac{\int_0^s f_1(u)du}{s} - C\right]ds.$$

The second term will vanish and the first term can be estimated by,

$$\left|\frac{1}{t}\int_{t-\theta}^t s\left[\frac{\int_0^s f_1(u)du}{s} - C\right]ds\right| \leq \int_{t-\theta}^t \left|\frac{\int_0^s f_1(u)du}{s} - C\right|ds.$$

The integrand on the right hand side converging to zero forces the entire integral to vanish and thus the theorem is proven. □

Remark 3 One should be careful when applying identity (19). We defined $f_1 := \frac{1}{\theta}\int_{t-\theta}^t f(s)ds$ with an extended version of f where $f(t) = 0$ for $t < 0$. This is equivalent to defining f_1 piecewise, i.e. $f_1(t) = \frac{1}{\theta}\int_0^t f(s)ds$ for $t \in [0, \theta]$ and $f_1(t) = \frac{1}{\theta}\int_{t-\theta}^t f(s)ds$ for $t > \theta$. This of course leads to a piecewise definition of $f_2 := f - f_1$, the reader should always keep this in mind as at first glance it may seem identity (19) is not correct for simple examples such as when $f(t) = c$ for $c \in \mathbb{R}\backslash\{0\}$. However with the consideration explained above one can calculate the explicit integrals and see the identity does indeed hold.

With this fundamental decomposition at hand we state and prove one more crucial lemma before providing our main result.

Lemma 4 *Let $\nu \in M(\mathbb{R}_+)$ and $g \in L^1(\mathbb{R}_+)$. Suppose $f \in \mathrm{Ces}(\mathbb{R}_+;\mathbb{R})$ with Cesàro limit, $C \in \mathbb{R}$. Then,*

$$\lim_{t\to\infty}\frac{1}{t}\int_0^t (f*\nu)(s)ds = C\nu(\mathbb{R}_+);\quad \lim_{t\to\infty}\frac{1}{t}\int_0^t (f*g)(s)ds = C\int_0^\infty g(s)ds.$$

Proof of Lemma 4 We only prove the statement for convolutions with finite measures, as the result for L^1 functions is a special case wherein the measure is absolutely continuous with respect to the Lebesgue measure. We show the difference $\left|\frac{1}{t}\int_0^t (f*\nu)(s)ds - C\nu(\mathbb{R}_+)\right|$, tends to zero as $t\to\infty$. Let $F(t) := \int_0^t f(s)ds$, then this difference (after an application of Fubini's theorem) becomes,

$$\left|\frac{1}{t}\int_{[0,t]} F(t-u)\nu(du) - C\nu(\mathbb{R}_+)\right|.$$

We first consider only the case when $|C|>0$ as the case when $C=0$ is simpler and follows from an analogous argument. Let $\varepsilon \in (0,1)$ be arbitrary and fix the positive constants K, T and α such that,

$$\left|\frac{1}{K}\int_0^K f(s)ds - C\right| < \frac{\varepsilon}{5|\nu|(\mathbb{R}_+)};\qquad |\nu|([T,\infty)) < \frac{\varepsilon}{5C|};$$

$$\alpha < \frac{\varepsilon}{5}\min\left\{\frac{1}{|C||\nu|(\mathbb{R}_+)}, \left(1-\frac{K}{K+1}\right)\right\}.$$

Then fix,

$$t > \max\left\{(K+1), (T+K), \frac{T}{\alpha}, \left(\frac{5|\nu|(\mathbb{R}_+)}{\varepsilon}\sup_{s\in[0,K]}\left|\int_0^s f(s)ds\right|\right)\right\}.$$

Now consider,

$$\begin{aligned}
&\left|\frac{1}{t}\int_{[0,t]} F(t-u)\nu(du) - C\nu(\mathbb{R}_+)\right| \\
&\quad\le \left|\frac{1}{t}\int_{[0,t-K]} F(t-u)\nu(du) - C\nu(\mathbb{R}_+)\right| + \frac{|\nu|(\mathbb{R}_+)}{t}\sup_{s\in[0,K]}\left|\int_0^s f(s)ds\right| \\
&\quad< \left|\frac{1}{t}\int_{[0,t-K]} F(t-u)\nu(du) - C\nu(\mathbb{R}_+)\right| + \frac{\varepsilon}{5}.
\end{aligned}$$

Next we make the substitution $s = t - u$,

$$\begin{aligned}
&\left|\frac{1}{t}\int_{[0,t-K]} F(t-u)\nu(du) - C\nu(\mathbb{R}_+)\right| + \frac{\varepsilon}{5} \\
&\quad = \left|\frac{1}{t}\int_{[K,t]} s\left[\frac{F(s)}{s} - C\right]\nu(ds) + \frac{C}{t}\int_{[K,t]} s\nu(ds) - C\nu(\mathbb{R}_+)\right| + \frac{\varepsilon}{5} \\
&\quad \le \left|\frac{C}{t}\int_{[K,t]} s\nu(ds) - C\nu(\mathbb{R}_+)\right| + \int_{[K,t]}\left|\frac{F(s)}{s} - C\right||\nu|(ds) + \frac{\varepsilon}{5} \\
&\quad < \left|\frac{C}{t}\int_{[K,t]} s\nu(ds) - C\nu(\mathbb{R}_+)\right| + \frac{\varepsilon|\nu|([K,t])}{5|\nu|(\mathbb{R}_+)} + \frac{\varepsilon}{5} \\
&\quad \le \left|\frac{C}{t}\int_{[K,t]} s\nu(ds) - C\nu(\mathbb{R}_+)\right| + \frac{2\varepsilon}{5}.
\end{aligned}$$

Now we undo the previous substitution and set $u = t - s$,

$$\begin{aligned}
&\left|\frac{C}{t}\int_{[K,t]} s\nu(ds) - C\nu(\mathbb{R}_+)\right| + \frac{2\varepsilon}{5} \\
&\quad = \left|\frac{C}{t}\int_{[0,t-K]} (t-u)\nu(du) - C\nu(\mathbb{R}_+)\right| + \frac{2\varepsilon}{5} \\
&\quad \le \left|C\nu([0,t-K]) - C\nu(\mathbb{R}_+)\right| + \left|\frac{C}{t}\int_{[0,t-K]} u\,\nu(du)\right| + \frac{2\varepsilon}{5} \\
&\quad < \frac{\varepsilon}{5} + \left|\frac{C}{t}\int_{[0,t-K]} u\,\nu(du)\right| + \frac{2\varepsilon}{5},
\end{aligned}$$

where the last inequality follows from the fact that $t > T + K$ and

$$|C\nu([0,t-K]) - C\nu(\mathbb{R}_+)| = |C||\nu|([t-K,\infty)).$$

Now our definition of α ensures $\alpha t < t - K$ when $t > K + 1$, thus we can write

$$\begin{aligned}
&\left|\frac{C}{t}\int_{[0,t-K]} u\,\nu(du)\right| + \frac{3\varepsilon}{5} \\
&\quad = \left|\frac{C}{t}\int_{[0,\alpha t]} u\,\nu(du) + \frac{C}{t}\int_{[\alpha t,t-K]} u\,\nu(du)\right| + \frac{3\varepsilon}{5} \\
&\quad \le \alpha\,|C||\nu|(\mathbb{R}_+) + \frac{|C|}{t}\int_{[\alpha t,t-K]} |u|\,|\nu|(du) + \frac{3\varepsilon}{5} \\
&\quad < \frac{\varepsilon}{5} + \frac{|C|}{t}\int_{[\alpha t,t]} |u|\,|\nu|(du) + \frac{3\varepsilon}{5} \\
&\quad < \frac{\varepsilon}{5} + |C||\nu|([\alpha t,\infty)) + \frac{3\varepsilon}{5}.
\end{aligned}$$

Now finally as $\alpha t > T$ we have, $|C||\nu|([\alpha t,\infty)) + \frac{4\varepsilon}{5} < \frac{\varepsilon}{5} + \frac{4\varepsilon}{5} = \varepsilon$. □

4 Results

We can now state and prove the main result of the paper.

Theorem 1 *Let x be the solution of equation* (10) *and assume $r \in L^1(\mathbb{R}_+;\mathbb{R})$. Then the following statements are true,*

(i) $\int_\cdot^{\cdot+\theta} f(s)ds \in \mathrm{Ces}(\mathbb{R}_+;\mathbb{R})$ *for all* $\theta \in (0,1] \iff x \in \mathrm{Ces}(\mathbb{R}_+;\mathbb{R})$.
(ii) $f \in \mathrm{Ces}(\mathbb{R}_+;\mathbb{R}) \iff x \in \mathrm{Ces}(\mathbb{R}_+;\mathbb{R})$ *and* $\dot{x} \in \mathrm{Ces}(\mathbb{R}_+;\mathbb{R})$.

Moreover in each case the limits are given explicitly, that is as $t \to \infty$,

(i)

$$\frac{1}{t}\int_0^t \int_s^{s+\theta} f(u)duds \longrightarrow C\theta; \qquad \frac{1}{t}\int_0^t x(s)ds \longrightarrow \frac{-C}{\nu(\mathbb{R}_+)}.$$

(ii)

$$\frac{1}{t}\int_0^t f(s)ds \longrightarrow C; \quad \frac{1}{t}\int_0^t x(s)ds \longrightarrow \frac{-C}{\nu(\mathbb{R}_+)}; \quad \frac{1}{t}\int_0^t \dot{x}(s)ds \longrightarrow 0.$$

where $C \in \mathbb{R}$.

Proof of Theorem 1 First we prove (i). Similarly to the proof of Theorem 1 in Appleby and Lawless (2023b), using the variation of constants formula (11), Lemma 3 and integrating by parts,

$$x(t) = r(t)\xi + (r * f_1)(t) + f_3(t) + (\dot{r} * f_3)(t), \quad t \geq 0,$$

where $f_3(t) := \int_0^t f_2(s)ds$. Thus we have,

$$\frac{1}{t}\int_0^t x(s)ds = \frac{1}{t}\int_0^t r(s)\xi ds + \frac{1}{t}\int_0^t (r * f_1)(s)ds + \frac{1}{t}\int_0^t f_3(s)ds + \frac{1}{t}\int_0^t (\dot{r} * f_3)(s)ds.$$

Sending $t \to \infty$ and applying Lemmas 3 and 4 we see,

$$\lim_{t\to\infty} \frac{1}{t}\int_0^t x(s)ds = \frac{-C}{\nu(\mathbb{R}_+)}.$$

For the converse we consider an extended solution so that $x(t) = 0$ for all $t < 0$, then we integrate (11) over the interval $[t-\theta, t]$ which results in,

$$\int_{t-\theta}^t f(u)du = x(t) - x(t-\theta) - \int_{t-\theta}^t \int_{[0,s]} x(s-u)\nu(du)ds.$$

Introducing the notation $X(t) := (x * \nu)(t)$ we obtain,

$$\frac{1}{t}\int_0^t \int_{s-\theta}^s f(u)duds = \frac{1}{t}\int_0^t x(s)ds - \frac{1}{t}\int_0^t x(s-\theta)ds - \frac{1}{t}\int_0^t \int_{s-\theta}^s X(u)duds.$$

When passing to the limit the first two terms on the RHS will cancel, thus we need only focus on the third. Now recall by Lemma 4,

$$\frac{1}{t}\int_0^t X(s)ds \longrightarrow \frac{-C}{\nu(\mathbb{R}_+)}\nu(\mathbb{R}_+) = -C,$$

as $t \to \infty$. Thus we may follow the proof of the converse of Lemma 3 with X in place of f_1 to obtain,

$$\lim_{t\to\infty} \frac{1}{t}\int_0^t \int_{s-\theta}^s X(u)duds = -\theta C.$$

But this implies,

$$\lim_{t\to\infty} \frac{1}{t}\int_0^t \int_{s-\theta}^s f(u)duds = \theta C,$$

as required. For the forward implication of (ii), recall Eq. (11), $x(t) = r(t)\xi + (r * f)(t)$. As $r \in L^1(\mathbb{R}_+; \mathbb{R})$ its Cesàro limit will be zero, thus,

$$\lim_{t\to\infty} \frac{1}{t}\int_0^t x(s)ds = \lim_{t\to\infty} \frac{1}{t}\int_0^t (r * f)(s)ds = \frac{-C}{\nu(\mathbb{R}_+)},$$

where the last equality follows from Lemmas 1 and 4. We now use this to show the Cesàro limit for $\dot{x}$. By (10) we obtain,

$$\frac{1}{t}\int_0^t \dot{x}(s)ds = \frac{1}{t}\int_0^t (x * \nu)(s)ds + \frac{1}{t}\int_0^t f(s)ds.$$

Passing to the limit and applying Lemma 4 and the limit just proven for x yields,

$$\frac{1}{t}\int_0^t \dot{x}(s)ds = \frac{-C}{\nu(\mathbb{R}_+)}\nu(\mathbb{R}_+) + C = 0.$$

For the converse, rearrange Eq. (10) to obtain $f(t) = \dot{x}(t) - \int_{[0,t]} x(t-s)\nu(ds)$. By supposition the Cesàro limit of $\dot{x}$ is zero and Lemma 4 tells us the Cesàro limit of the convolution is $\frac{C}{\nu(\mathbb{R}_+)}\nu(\mathbb{R}_+) = C$, as required. □

The assumptions that $r, r_\tau \in L^1(\mathbb{R}_+; \mathbb{R})$ are known to be sharp in order to obtain admissibility results for solutions to equations (10) and (13). We provide a converse result for the finite memory problem which shows that $r_\tau \in L^1(\mathbb{R}_+; \mathbb{R})$ is a necessary condition in order for solutions of (13) to admit a Cesàro limit. Throughout the proof we make heavy use of the semi-explicit asymptotic expansion of the resolvent, this approach is much more involved in the Volterra case due to the lack of information about the general form of the Volterra resolvent. It is in this regard that we only consider the finite memory equation.

Proposition 1 *Let x be the solution of* (13). *If for each $\psi \in C([-\tau, 0]; \mathbb{R})$ and $f \in L^1_{loc}(\mathbb{R}_+; \mathbb{R})$ such that $t \mapsto \int_t^{t+\theta} f(s)ds \in \text{Ces}(\mathbb{R}_+; \mathbb{R})$ for all $\theta \in (0, 1]$, we have $x(\cdot, \psi) \in \text{Ces}(\mathbb{R}_+; \mathbb{R})$, then $r_\tau \in L^1(\mathbb{R}_+; \mathbb{R})$.*

For the readers convenience, before we provide a proof of Theorem 1 we recall some common notation in the field of functional equations. Let,

$$x_0(t, \psi) := r_\tau(t)\psi(0) + \int_{[-\tau,0]} \left(\int_s^0 r_\tau(t + s - u)\psi(u)du \right) \mu(ds), \tag{20}$$

so we can rewrite (16) as $x(t, \psi) = x_0(t, \psi) + (r_\tau * f)(t)$. Thus x_0 solves Eq. (13) with the perturbation term switched off. The roots of the transcendental equation, $h(\lambda) = \lambda - \int_{[-\tau,0]} e^{-\lambda s}\mu(ds)$, govern the asymptotic behaviour of r_τ. Let $\Lambda = \{\lambda \in \mathbb{C} : h(\lambda) = 0\}$ and define $v_0(\mu) := \sup\{\text{Re}(\lambda) : \lambda \in \Lambda\}$. The cardinality of $\Lambda' = \{\lambda \in \Lambda : \text{Re}(\lambda) = v_0(\mu)\}$ is finite, thus r_τ admits the expansion,

$$e^{-v_0(\mu)t} r_\tau(t) = \sum_{\lambda_j \in \Lambda'} \left\{ p_j(t)\cos(\beta_j t) + q_j(t)\sin(\beta_j t) \right\} + g(t), \tag{21}$$

where p_j and q_j are polynomials of degree $m_j - 1$ (where m_j is the multiplicity of λ_j), $\beta_j = \text{Im}(\lambda_j)$ and $g(t) = o(e^{-\varepsilon t})$ for some $\varepsilon > 0$. For proofs of all facts recalled above the reader may consult (Diekmann et al. 1995, Chap. I).

Proof of Theorem 1 We aim to prove $v_0(\mu) < 0$ which yields automatically $r_\tau \in L^1(\mathbb{R}_+; \mathbb{R})$ due to the expansion (21). We proceed by contradiction, thus assume $v_0(\mu) > 0$. Setting $f = 0$ yields $x_0(\cdot, \psi) \in \text{Ces}(\mathbb{R}_+; \mathbb{R})$ for all $\psi \in C([-\tau, 0]; \mathbb{R})$. Hence select $\lambda \in \Lambda'$ and set $\psi = \text{Re}(e^{\lambda t})$ which yields the solution, $x_0(t; \psi) = \text{Re}(e^{\lambda t})$ for all $t \in [-\tau, \infty)$. Let $\lambda = v_0(\mu) + i\beta$, then

$$\int_0^t \text{Re}(e^{\lambda s})ds = \int_0^t e^{v_0(\mu)s}\cos(\beta s)ds = \frac{e^{v_0(\mu)t}(\beta\sin(v_0(\mu)t) + v_0(\mu)\cos(\beta t))}{|\lambda|^2},$$

which upon dividing by t will not tend to a limit, contradicting $x_0(\cdot, \psi) \in \text{Ces}(\mathbb{R}_+; \mathbb{R})$. Thus we must have $v_0(\mu) \le 0$. Now assume $v_0(\mu) = 0$, in which we distinguish two cases. First assume there exists a $\lambda \in \Lambda'$ with multiplicity strictly great than one. In this case we fix $\psi = t\cos(\text{Im}(\lambda)t)$ which again yields a solution $x_0(t; \psi) = t\cos(\text{Im}(\lambda)t)$ for all $t \in [-\tau, \infty)$, but as before a simple calculation shows

$$\frac{1}{t}\int_0^t s\cos(\text{Im}(\lambda)s)ds = \frac{\sin(\text{Im}(\lambda)t)}{\text{Im}(\lambda)} + \frac{\cos(\text{Im}(\lambda)t) - 1}{t\text{Im}(\lambda)^2},$$

which oscillates indefinitely and so $x_0(\cdot;\psi) \notin \mathrm{Ces}(\mathbb{R}_+;\mathbb{R})$, a contradiction. Thus it must be when $v_0(\mu) = 0$, all elements of Λ' have multiplicity one. In this case,

$$r_\tau(t) = \sum_{j=1}^{n} \left\{c_j \cos(\beta_j t) + k_j \sin(\beta_j t)\right\} + g(t),$$

where $c_j, k_j \in \mathbb{R} \setminus \{0\}$. Now we fix the initial condition $\psi = 0$ which yields, $x(t, 0) = (r_\tau * f)(t)$. Now assume first that $n > 1$ which means at least one zero is not at the origin. Without loss of generality we may assume $\beta_1 \neq 0$. Choose $f(t) = k_1 \sin(\beta_1 t) - c_1 \cos(\beta_1 t)$ and define $F(t) := \int_0^t f(s)ds$, we see $\int_0^t x(u, 0)du = \int_0^t r_\tau(u)F(t-u)du$ with,

$$F(t) = \frac{c_1}{\beta_1}\cos(\beta_1 t) + \frac{k_1}{\beta_1}\sin(\beta_1 t) - \frac{c_1}{\beta_1}.$$

Now we can calculate $\int_0^t x(u, 0)du$ semi-explicitly,

$$\begin{aligned}
&\frac{1}{t}\int_0^t r_\tau(u)F(t-u)du \\
&= \sin(\beta_1 t)\left(\frac{c_1 k_1}{\beta_1} + \frac{(c_1^s - k_1^2)}{2t\beta_1^2}\right) + \cos(\beta_1 t)\left(\frac{c_1 k_1}{t\beta_1^2} + \frac{(c_1^2 - k_1^2)}{2\beta_1}\right) - \frac{c_1 k_1}{t\beta_1^2} \\
&\quad + \frac{1}{t}\sum_{j=1}^{n}\left\{A_{1,j}\cos(\beta_j t) + A_{2,j}\sin(\beta_j t) + A_{3,j}\cos(\beta_1 t) + A_{4,j}\sin(\beta_1 t) + A_{5,j}\right\} \\
&\quad + \frac{1}{t}\int_0^t g(u)F(t-u)du,
\end{aligned}$$

where $A_{i,j} \in \mathbb{R}$. The summation term will vanish as $t \to \infty$ and the exponential estimate on g ensures the integral term will also vanish. However we see the first term will oscillate indefinitely as the constants c_1, k_1 and β_1 are non-zero. This means $x(\cdot; 0) \notin \mathrm{Ces}(\mathbb{R}_+;\mathbb{R})$ which is a contradiction. Now if n=1 there is only a single root on the imaginary axis. If this root is not at the origin then the above argument still holds, so we assume the sole root is at the origin. In this case $r_\tau(t) = c + g(t)$ where $c \in \mathbb{R}\setminus\{0\}$. Now set $\psi = 0$ and $f(t) = 1/c$, then $x(t, 0) = t + (g * f)(t)$. As $g \in L^1(\mathbb{R}_+;\mathbb{R})$ and $f \in \mathrm{Ces}(\mathbb{R}_+;\mathbb{R})$, Lemma 4 implies the convolution has a Cesáro limit but the linear function $f(t) = t$ will not, which yields the final contradiction. Thus we must have that $v_0(\mu) < 0$ and the claim is proven. □

Proposition 1, Theorem 1 and Remark 1, provide the proof of our next theorem.

Theorem 2 *Let $x(\cdot,\psi)$ be the solution to* (13).
(A) Suppose $r_\tau \in L^1(\mathbb{R}_+;\mathbb{R})$, then for all $\psi \in C([-\tau, 0];\mathbb{R})$ the following hold true,

(i) $\int_\cdot^{\cdot+\theta} f(s)ds \in \mathrm{Ces}(\mathbb{R}_+;\mathbb{R})$ *for all* $\theta \in (0, 1] \iff x(\cdot,\psi) \in \mathrm{Ces}(\mathbb{R}_+;\mathbb{R})$.
(ii) $f \in \mathrm{Ces}(\mathbb{R}_+;\mathbb{R}) \iff x(\cdot,\psi) \in \mathrm{Ces}(\mathbb{R}_+;\mathbb{R})$ *and* $\dot{x}(\cdot,\psi) \in \mathrm{Ces}(\mathbb{R}_+;\mathbb{R})$.

Moreover in each case the limits are given explicitly, that is as $t \to \infty$,

(i)

$$\frac{1}{t}\int_0^t \int_s^{s+\theta} f(u)duds \to C\theta; \qquad \frac{1}{t}\int_0^t x(s,\psi)ds \to \frac{-C}{\mu([-\tau,0])}.$$

(ii)

$$\frac{1}{t}\int_0^t f(s)ds \to C; \quad \frac{1}{t}\int_0^t x(s,\psi)ds \to \frac{-C}{\mu([-\tau,0])}; \quad \frac{1}{t}\int_0^t \dot{x}(s,\psi)ds \to 0.$$

where $C \in \mathbb{R}$.

*(**B**) The following are equivalent,*

(i) $r_\tau \in L^1(\mathbb{R}_+;\mathbb{R})$.

(ii) For all $\psi \in C([-\tau,0];\mathbb{R})$ *and* $f \in \mathrm{Ces}(\mathbb{R}_+;\mathbb{R})$ *we have* $x(\cdot,\psi) \in \mathrm{Ces}(\mathbb{R}_+;\mathbb{R})$.

(iii) For all $\psi \in C([-\tau,0];\mathbb{R})$ *and* f *s.t.* $\int_\cdot^{\cdot+\theta} f(s)ds \in \mathrm{Ces}(\mathbb{R}_+;\mathbb{R})$ *for all* $\theta \in (0,1]$ *we have* $x(\cdot,\psi) \in \mathrm{Ces}(\mathbb{R}_+;\mathbb{R})$.

5 Example

In this section we show the statements $f \in \mathrm{Ces}(\mathbb{R}_+;\mathbb{R})$ and $t \mapsto \int_t^{t+\theta} f(s)ds \in \mathrm{Ces}(\mathbb{R}_+;\mathbb{R})$ for each $\theta \in (0,1]$ are in fact not equivalent. In general the first implies the second, to see this recall that if $f \in \mathrm{Ces}(\mathbb{R}_+;\mathbb{R})$ then by Theorem 1 part (ii), we necessarily have solutions of (10) obeying $x \in \mathrm{Ces}(\mathbb{R}_+;\mathbb{R})$. But then applying part (i) of Theorem 1 yields $t \mapsto \int_t^{t+\theta} f(s)ds \in \mathrm{Ces}(\mathbb{R}_+;\mathbb{R})$ for each $\theta \in (0,1]$. This argument, although mathematically sound, is unsatisfactory due to its indirect nature. Indeed this can be proven in a direct fashion with an apt application of Lemma 4. Let $f \in \mathrm{Ces}(\mathbb{R}_+;\mathbb{R})$ and introduce the function $g(t) = \theta^{-1}\chi_{\{t\in[0,\theta]\}}(t)$. Then $g \in L^1(\mathbb{R}_+;\mathbb{R})$, hence by Lemma 4 $(f * g) \in \mathrm{Ces}(\mathbb{R}_+;\mathbb{R})$, but for $t > \theta$,

$$(f * g)(t) = \int_0^\theta g(s)f(t-s)ds = \frac{1}{\theta}\int_{t-\theta}^t f(s)ds.$$

Thus we have proven as claimed that,

$$f \in \mathrm{Ces}(\mathbb{R}_+;\mathbb{R}) \implies t \mapsto \int_t^{t+\theta} f(s)ds \in \mathrm{Ces}(\mathbb{R}_+;\mathbb{R}) \text{ for all } \theta \in (0,1]. \quad (22)$$

We state a counter example that shows the reverse implication is in general false, consider $f : \mathbb{R}_+ \to \mathbb{R} : t \mapsto t^{\alpha+1}\sin(t^{\alpha+1}) \quad \alpha > 0$. This claim can be verified after some tedious but elementary calculations. Although the reverse implication of (22) is not true in general we provide a side condition which yields the converse in the case that the function is positive.

Proposition 2 *Let $f \in L^1_{loc}(\mathbb{R}_+; \mathbb{R}_+)$. The following are equivalent,*

$$(i)\ \lim_{t\to\infty}\frac{1}{t}\int_0^t\int_s^{s+\theta} f(u)duds = C\theta \text{ for all } \theta \in (0,1];\quad \lim_{t\to\infty}\frac{1}{t}\int_t^{t+1} f(s)ds = 0.$$

$$(ii)\ \lim_{t\to\infty}\frac{1}{t}\int_0^t f(s)ds = C.$$

Proof of Proposition 2 The fact that (ii) implies (i) is clear from the discussion at the beginning of this section and from the identity,

$$\frac{1}{t}\int_t^{t+1} f(s)ds = \frac{1}{t}\int_0^{t+1} f(s)ds - \frac{1}{t}\int_0^t f(s)ds.$$

To see the other implication we write for $t > \theta$,

$$\frac{1}{t}\int_0^t\int_s^{s+\theta} f(u)duds \tag{23}$$

$$= \frac{1}{t}\int_0^\theta f(u)(u-\theta)du + \frac{\theta}{t}\int_0^t f(u)du + \frac{1}{t}\int_t^{t+\theta} f(u)(t+\theta-u)du.$$

The first and third term on the right hand side will vanish upon sending $t \to \infty$ and thus we obtain the desired limit. □

6 Comparison with Integral Equations

To this point we have considered integro-differential and functional equations and have shown solutions to perturbed equations may admit a Cesàro limit even when the perturbation function does not. This phenomenon is impossible for integral equations, in this section we provide a proof. Consider the integral equation,

$$x(t) = \int_0^t k(t-s)x(s)ds + f(t), \quad t \geq 0, \tag{24}$$

where $f \in L^1_{loc}(\mathbb{R}_+; \mathbb{R})$ and $k \in L^1(\mathbb{R}_+; \mathbb{R})$. In this regime the integral resolvent of k denoted by r_k is the solution to the convolution equation $r_k = k + r_k * k$, on the halfline $\mathbb{R}_+$. The standard variation of constants formula yields, $x(t) = f(t) + (r_k * f)(t), t \geq 0$. Now let $r_k \in L^1(\mathbb{R}_+; \mathbb{R})$. Assume $f \in \text{Ces}(\mathbb{R}_+; \mathbb{R})$ then Lemma 4 implies $x \in \text{Ces}(\mathbb{R}_+; \mathbb{R})$. Conversely assume $x \in \text{Ces}(\mathbb{R}_+; \mathbb{R})$, then with $k \in L^1(\mathbb{R}_+; \mathbb{R})$ we can rearrange (24) and apply Lemma 4 to give $f \in \text{Ces}(\mathbb{R}_+; \mathbb{R})$.

7 Conclusion

In this paper we have characterised when the solution of a perturbed linear functional differential equation lies in the space $\text{Ces}(\mathbb{R}_+; \mathbb{R})$ and also provided necessary and sufficient conditions to ensure the solution of a perturbed linear integro-differential Volterra equation lies in $\text{Ces}(\mathbb{R}_+; \mathbb{R})$.[4] It is interesting to reflect on the improvements made with respect to the classical theory. For classical admissibility theory, three things are needed to ensure solutions of (10) reside in a particular space V while operating under the assumptions that $r \in L^1(\mathbb{R}_+; \mathbb{R})$ and $\nu \in M(\mathbb{R}_+; \mathbb{R})$, namely:

1. The resolvent obeys, $r \in V$.
2. If $g \in L^1(\mathbb{R}_+; \mathbb{R})$ then the operator $h \mapsto g * h$ must map V into itself.
3. The perturbation function obeys $f \in V$.

This paper along with Appleby and Lawless (2023b) have shown that the third point can be significantly weakened. We can have $f \notin V$ and yet solutions still reside in V. This is contingent on the perturbation function admitting a decomposition of the form $f = f_1 + f_2$ such that $f_1 \in V$ and $t \mapsto \int_0^t f_2(s)ds \in V$. Even after scrutiny of the proofs of Lemmas 2 and 3 it seems unclear how one may generally describe such spaces. However even with this said, it is the authors opinion that given any "*reasonable*" space that is usually considered for admissibility theory that satisfies point 2 above, a decomposition result like the one described can be proven. Following the authors recent works (Appleby and Lawless 2023a, b, 2024) there is mounting evidence to suggest that in order to obtain sharp results for perturbed dynamical systems (deterministic or stochastic) the correct condition to study is (4). An interesting extension would be to consider emulating such methodology for infinite dimensional equations like those found in Prüss (1993).

Acknowledgements The authors wish to thank Prof. Elena Braverman who organised the minisymposium on "Delay and stochastic differential equations in life sciences and engineering" at the ICIAM confererence 2023, at which an early version of this work was presented.
The authors also wish to acknowledge the careful reading of our manuscript by the anonymous referees.
EL is supported by Science Foundation Ireland (16/IA/4443). JA is supported by the RSE Saltire Facilitation Network on Stochastic Differential Equations: Theory, Numerics and Applications (RSE1832). EL would like to thank Ole Cañadas for insightful discussions resulting in the improvement of this work.

References

Appleby, J.A.D.: Almost sure stability of linear Itô-Volterra equations with damped stochastic perturbations. Electron. Commun. Probab. **7**, 223–234 (2002)

[4] under the assumption of an integrable resolvent.

Appleby, J.A.D.: Subexponential solutions of linear Itô-Volterra equations with a damped perturbation. **11**, 5–10 (2004)
Appleby, J.A.D.: Mean square characterisation of a stochastic Volterra integrodifferential equation with delay. Int. J. Dyn. Syst. Differ. Equ. **11**(3–4), 194–226 (2021)
Appleby, J.A.D., Freeman, A.: Exponential asymptotic stability of linear Itô-Volterra equations with damped stochastic perturbations. Electron. J. Probab. **8**(22) (2003)
Appleby, J.A.D., Lawless, E.: Mean square asymptotic stability characterisation of perturbed linear stochastic functional differential equations. Appl. Numer. Math. (2023)
Appleby, J.A.D., Lawless, E.: Solution space characterisation of perturbed linear volterra integrodifferential convolution equations: The L^p case. Appl. Math. Lett. **146**, 108825 (2023)
Appleby, J.A.D., Lawless, E.: Solution space characterisation of perturbed linear discrete and continuous stochastic volterra convolution equations: the ℓ^p and L^p cases (2024). arxiv:2407.07767
Appleby, J.A.D., Patterson, D.D.: On the admissibility of unboundedness properties of forced deterministic and stochastic sublinear Volterra summation equations. Electron. J. Qual. Theory Differ. Equ. Paper No. 63, 44 (2016)
Appleby, J.A.D., Patterson, D.D.: Large fluctuations and growth rates of linear volterra summation equations. Differ. Eqn. Appl. **23**(6), 1047–1080 (2017)
Appleby, J.A.D., Riedle, M.: Almost sure asymptotic stability of stochastic Volterra integro-differential equations with fading perturbations. Stoch. Anal. Appl. **24**(4), 813–826 (2006)
Appleby, J.A.D., Mao, X., Riedle, M.: Geometric Brownian motion with delay: mean square characterisation. Proc. Am. Math. Soc. **137**(1), 339–348 (2009)
Beneš, V.E.: A nonlinear integral equation in the Marcinkiewicz space m_2. J. Math. Phys. **44**, 24–35 (1965)
Berger, M.A., Mizel, V.J.: Volterra equations with Itô integrals. I. J. Integr. Eqn. **2**(3), 187–245 (1980)
Berger, M.A., Mizel, V.J.: Volterra equations with Itô integrals. II. J. Integr. Eqn. **2**(4), 319–337 (1980)
Bertrandias, J.P.: Espaces de fonctions bornées et continues en moyenne asymptotique d'ordre p. Bull. Soc. Math. France Mém. **5**, 106 (1966)
Bingham, N.H., Goldie, C.M., Teugels, J.L.: Regular variation. Encyclopedia of Mathematics and Its Applications, vol. 27. Cambridge University Press, Cambridge (1989)
Boykov, I.V., Roudnev, V.A., Boykova, A.I.: Stability of solutions of systems of Volterra integral equations. Appl. Math. Comput. **475**, Paper No. 128728 (2024)
An introduction to theory and applications. In: Volterra Integral Equations. Cambridge Monographs on Applied and Computational Mathematics, vol. 30. Cambridge University Press, Cambridge (2017)
Chan, T., Williams, D.: An "excursion" approach to an annealing problem. Math. Proc. Cambridge Philos. Soc. **105**(1), 169–176 (1989)
Corduneanu, C.: Integral Equations and Stability of Feedback Systems. Academic, New York (1973)
Corduneanu, C.: Integral Equations and Applications. Cambridge University Press, New York (1990)
Corduneanu, C.: Almost Periodic Oscillations and Waves. Springer, New York (2009)
Diekmann, O., van Gils, S.A., Verduyn Lunel, S.M., Walther, H.-O.: Functional, complex, and nonlinear analysis. In: Delay Equations. Applied Mathematical Sciences, vol. 110. Springer, New York (1995)
Gikhman, I.I., Skorokhod, A.V.: Stochastic Differential Equations. Springer, New York, Heidelberg (1972)
Gripenberg, G., Londen, S.-O., Staffans, O.: Volterra integral and functional equations. In: Encyclopedia of Mathematics and It's Applications. Cambridge University Press (1990)
Mizel, V.J., Trutzer, V.: Stochastic hereditary equations: existence and asymptotic stability. J. Integr. Eqn. **7**(1), 1–72 (1984)
Pardoux, É., Protter, P.: Stochastic Volterra equations with anticipating coefficients. Ann. Probab. **18**(4), 1635–1655 (1990)

Protter, P.: Volterra equations driven by semimartingales. Ann. Probab. **13**(2), 519–530 (1985)

Prüss, J.: Evolutionary integral equations and applications. Monographs in Mathematics, vol. 87. Birkhäuser Verlag, Basel (1993)

Strauss, A., Yorke, J.A.: Perturbation theorems for ordinary differential equations. J. Differ. Eqn. **3**, 15–30 (1967)

Strauss, A., Yorke, J.A.: On asymptotically autonomous differential equations. Math. Syst. Theory **1**, 175–182 (1967)

Stochastic Modelling and Applications of Big Data in Finance

Anatoliy Swishchuk

Abstract In this paper we give an overview of big data modelling and applications in finance. The models include semi-Markov, general semi-Markov, Hawkes, non-linear Hawkes, general compound Hawkes, exponential general compound Hawkes and multivariate general compound Hawkes models. Applications are given to real big data such as LOBster (5 stocks, 21/06/2012), Deutsche Boerse Group (15 stocks, 23/09/2013), 5 stocks data from Cartea et al. book and CISCO data (3/11/2014).

Keywords Big data · Limit order book · Stochastic modelling · Hawkes process · Semi-Markov process · Finance

1 Introduction

Big data has now become a driver of model building and analysis in a number of areas, including finance. (Big Data may be defined as data that contains greater Variety, arriving in increasing Volume, and with more Velocity. This is also known as the three Vs). There are hundred thousands of orders for one stock just for one day: e.g., for CISCO data on Nov 3, 2014, we have 0.5 million price orders (!) for that day, and that is only for 1-level order book, meaning the limit orders sitting at the best bid and ask. And if we take hundreds of stocks on an exchange and not only 1-level orders book, then we will get an example of really big data in finance! And the question is: can we model this mechanism of trading and describe this big data in finance? And the answer is 'Yes'. Below we show and explain how to do this for LOBster and other real big data. Main problem: how to deal with big data arising in electronic markets for algorithmic and high-frequency (milliseconds) trading that contain two types of orders, limit orders and market orders. More than half of the markets in today's highly competitive and relentlessly fast-paced financial world now use a limit order book (LOB) mechanism to facilitate trade. Let us describe this

A. Swishchuk (✉)
University of Calgary, Alberta, Canada
e-mail: aswish@ucalgary.ca

E. Braverman and A. Swishchuk (eds.), *Delay and Stochastic Differential Equations*, ICIAM2023 Springer Series 12, https://doi.org/10.1007/978-981-95-7310-3_5

mechanism in brief. Orders to buy and sell an asset arrive at an exchange: (1) market buy/sell order—specifies number of shares to be bought/sold at the best available price, right away; (2) limit buy/sell order—specifies a price and a number of shares to be bought/sold at that price, when possible; (3) order cancellation—agents who have submitted a limit order may cancel the order before it is executed; market orders are executed immediately; limit orders are queued for later execution, but may cancel. Thus, the Limit-Order Book (LOB) is the collection of queued limit orders awaiting execution or cancellation. Usually, we are interested in mid-price of a stock: $S_t := \frac{s_t^a + s_t^b}{2}$, where $s_t^i, i = a, b$, are ask- (the lowest limit sell order price in the book) or bid-price (the highest limit buy order price in the book). Numerical results, associated with Lobster and other data, are presented, and explanation and justification of our method of studying of big data in finance are considered. We also describe our stochastic models for mid-price changes in the market, and explain how to use this model in practice. Thus, in this paper we give an overview of big data modelling and applications in finance. The models include semi-Markov, general semi-Markov, Hawkes non-linear Hawkes, general compound Hawkes, exponential general compound Hawkes and multivariate general compound Hawkes models. Applications are given to real big data such as LOBster (5 stocks, 21/06/2012), Deutsche Boerse Group (15 stocks, 23/09/2013), 5 stocks data from Cartea et al. book, and CISCO data (3/11/2014).

The paper is organized as follows. Big data modelling in finance is described in Sect. 2. Section 3 presents big data applications in finance. In Sect. 4 we consider some numerical examples.

2 Big Data Modellig in Finance

In this section we describe different big data models in finance such as semi-Markov, general semi-Markov, compound Hawkes, general compound Hawkes, non-linear Hawkes, exponential general compound Hawkes and multivariate Hawkes models.

2.1 Semi-Markov Model

For the semi-Markov model (see Swishchuk and Vadori (2017)) we describe the state process, which is semi-Markov process, and which will keep track of the state of the limit order book at time t (stock price and sizes of the bid and ask queues), $L_t := (S_t, q_t^b, q_t^a)$, where $S_t := (s_t^a + s_t^b)/2$ is a mid-price, $S_t := s_0 + \sum_{i=1}^{N(t)} X_k$, $X_k = \{-\delta, +\delta\}$, δ-tick size, q_t^a, q_t^b are sizes of bid and ask queues, $N(t)$-number of price changes (renewal process). We note, that if we "add" to this process the process (V_t^b, V_t^a) keeping track of the nature of the last book event at the bid and the ask to make it Markovian: in this sense we can view it as being semi-Markovian.

The process:$L_t := (S_t, q_t^b, q_t^a, V_t^b, V_t^a)$ is Markovian, where V_t^b, V_t^a are processes for events of increase or decrease the bid or ask queue by 1, respectively.

2.2 *General Semi-Markov Model*

The paper (Swishchuk et al. 2017) considers a general semi-Markov model for Limit Order Books with two states, which incorporates price changes that are not fixed to one tick. Furthermore, we introduced an even more general case of the semi-Markov model for limit order books that incorporates an arbitrary number of states (controlled by X_k) for the price changes: $S_t := s_0 + \sum_{i=1}^{N(t)} a(X_k)$. We call it general semi-Markov model for the LOB, because of $a(x)$—a bounded function of the states of Markov chain X_k, $N(t)$ here is renewal process.

2.3 *Compound and Regime-Switching Compound Hawkes Models*

Trading activity, which is not completely memoryless process, leads to time series of irregularly spaced points that show a clustering behaviour. This stylized property suggests the use of the Hawkes process (HP), which is an extension of the classical Poisson process that is not suitable for modelling trade arrival times. The evidence of this is, e.g., a QQ-plot of inter-arrival times of trades against an exponential distribution for Eurostoxx futures trades on March 03, 2011, which rejects the Poisson process as the data-generating process for the order flow. (See, e.g., Da- Fonseca and Zaatour (2014), The Journal of Futures Markets, v. 34, No. 6, pp. 548–579). Trades tend actually to cluster: histogram of the number of trades occurring every minute during a trading day for the Eu- rostoxx shows this feature (see, e.g., DaFonseca and Zaatour (2014), The Journal of Futures Markets, v. 34, No. 6, pp. 548–579) (Fonseca and Zaatour 2013). The Hawkes process (HP) is so-called self-exciting point process which means that it is a point process with a stochastic intensity which, through its dependence on the history of the process, captures the temporal and cross sectional dependence of the event arrival process, as well as the self-exciting property observed in empirical analysis (see Hawkes (1971) and Hawkes (1971)). The one-dimensional Hawkes process is a point process $N(t)$ which is characterized by its intensity $\lambda(t)$ with respect to its natural filtration: $\lambda(t) = \lambda + \int_0^t \mu(t-s)dN(s)$, where $\lambda > 0$ is a background intensity, and the response/excitation function $\mu(t)$ is a positive function and satisfies $\int_0^{+\infty} \mu(s)ds < 1$.

In the paper (Swishchuk et al. 2019) we introduced two new Hawkes models, namely, compound and regime-switching compound Hawkes models, to model the price processes in the limit order books: $S_t := s_0 + \sum_{i=1}^{N(t)} X_k$, where $N(t)$ is a

Hawkes process. We note, that X_k takes here also only two values: $X_k \in \{+\delta, -\delta\}$, thus X_k is a two-state Markov chain.

2.4 General Compound Hawkes Model (GCHM)

Let $N(t)$ be an one-dimensional Hawkes process defined above. Let also X_n be ergodic discrete-time finite (or possibly infinite but countable) state Markov chain, independent of $N(t)$, with space state X, and $a(x)$ be any bounded function on X. The general compound Hawkes process is defined as $S_t = S_0 + \sum_{k=1}^{N(t)} a(X_k)$. The general compound Hawkes process was first introduced in Sw. (2017) (IME 2017, Vienna) (Swishchuk 2017) to model the risk process in insurance, and studied in detailed in Sw. 2018 (Wilmott Magazine, v. 2018, Issue 94, pp. 50–57) (Swishchuk 2018). The GCHM is general enough to include: compound Poisson model: $S_t = S_0 + \sum_{k=1}^{N(t)} X_k$, where $N(t)$ is a Poisson process and $a(X_k) = X_k$ are i.i.d.r.v.; compound Hawkes model, introduced above: $S_t = S_0 + \sum_{k=1}^{N(t)} X_k$, where $N(t)$ is a Hawkes process and $a(X_k) = X_k$ are i.i.d.r.v.; and semi-Markov and general semi-Markov models introduced above: $S_t = S_0 + \sum_{k=1}^{N(t)} a(X_k)$, where $N(t)$ is a renewal process and X_k is a Markov chain.

2.5 Non-Linear General Compound Hawkes Model

Non-linear general compound Hawkes model is defined as $S_t = S_0 + \sum_{i=1}^{N_t} a(X_k)$, where $X_k \in \{1, 2, ..., n\} := X$ is a discrete-time Markov chain, $a(x)$ is a bounded function on $X = \{1, 2, ..., n\}$, $N(t)$ is the the non-linear Hawkes process defined by the intensity function in the following form $\lambda(t) = h\Big(\lambda + \int_0^t \mu(t-s)dN(s)\Big)$, with $h(\cdot)$ being a non-linear increasing function with support in R^+, α-Lipschitz and such that $\alpha||h||_{L^1} < 1$. See Swishchuk and Huffman (2020) for more details.

2.6 Exponential General Compound Hawkes Model

The exponential general compound Hawkes process (EGCHP) is defined as follows:

$$S_t = S_0 \exp\left(\sum_{k=1}^{N_t} X_k\right),$$

where N_t is one-dimensional Hawkes process, and X_k is n-state Markov chain.

2.7 Multivariate Hawkes Model

Define $S_{it} = S_{i0} + \sum_{k=1}^{N_{i,t}} a(X_{ik}), \quad t \geq 0$, where $\mathbf{N}_t = (N_{1t}, \ldots, N_{dt})$ is a simple multivariate point process (e.g., Hawkes process, etc.) with intensity $\boldsymbol{\lambda}_t = (\lambda_{1t}, \ldots, \lambda_{dt})$, $X_{i,k}$ is a N-state Markov chain for each i, $i = 1, 2, ..., d$, $a(x)$ is a bounded function on $X = \{1, 2, ..., N\}$. Here: $\lambda_{it} = \lambda_i + \sum_{j=1}^{d} \int_{(0,t)} \mu_{ij}(t - s)dN_{j,s}$, $i \in \{1, \ldots, d\}$, which can be written in vector form as $\boldsymbol{\lambda}_t = \boldsymbol{\lambda} + \int_{(0,t)} \mu(t - s)d\mathbf{N}_s$, where $\boldsymbol{\lambda}_t \in \mathbf{R}_+^d = [0, \infty)^d$ is bounded, $\mu(t) \in \mathbf{R}_+^{d \times d}$. Further set $\mathbf{S}_t = (S_{1,t}, \ldots, S_{d,t})$. We call this vector stochastic process as multivariate general compound Hawkes model. If we take $d = 5$, for example, then $\mathbf{S}_t = (S_{1,t}, S_{2,t}, ..., S_{5,t})$ could be a vector for 5 stocks in Lobster data or for 5 stocks from Cartea et al. (2015) book.

3 Big Data Applications in Finance

3.1 Application Od Semi-Markov Model See Sect. 2.1

As suggested by empirical observations (LOBster data), we extend Cont and de Larrard (2013) (SIAM J. Finan. Math, 2013) framework to: (1) arbitrary distributions for book events inter-arrival times (non-exponential) and (2) both the nature of a new book event and its corresponding inter-arrival time depend on the nature of the previous book event (non-independent).

3.2 Application of General Semi-Markov Model See Sect. 2.2

For both cases the justification, diffusion limits, where we used martingale approach (see Swishchuk and Vadori (2017)), implementations and numerical results are presented for different limit order book data: Apple, Amazon, Google, Microsoft, Intel on 2012/06/21 and Cisco, Facebook, Intel, Liberty Global, Liberty Interactive, Microsoft, Vodafone from 2014/11/03 to 2014/11/07.

3.3 Application of Compound and Regime-Switching Compound Hawkes Models See Sect. 2.3

Compound and regime-switching compound Hawkes models were applied to CISCO data (5 days, 3-7 November 2014). The functional central limit theorems, considered in the paper (Swishchuk et al. 2019), provides insights into the link between price volatilities and order flows in limit order books with several assets. This important result relates the volatility of intraday returns at lower frequencies to the high-frequency arrival rates of orders. The numerical example shows that error of estimation is approximately 0.08, indicating that approximation in diffusion limit for compound Hawkes model is pretty good. Similar result is valid for regime-switching compound Hawkes model.

3.4 Application of General Compound Hawkes Model See Sect. 2.4

We study various new Hawkes processes, namely, so-called general compound and regime-switching general compound Hawkes processes, and apply them to model the price processes in the limit order books. We prove Law of Large Numbers (LLN) and Functional Central Limit Theorems (FCLT) for these models. The latter two FCLTs are applied to limit order books where we use these asymptotic methods to study the link between price volatility and order flow in our two models by investigating the diffusion limits of these price processes. The volatilities of price changes are expressed in terms of parameters describing the arrival rates and price changes. Quantitative and comparative analyses of limit order books with general compound Hawkes model is given in Swishchuk and He (2019). Numerical analysis is applied to the following data: EBAY: eBay Inc.; FB: Facebook Inc.; MU: MicronTechnology Inc. PCAR: PACCAR Inc.; SMH: VANECK Inc. which can be found in Cartea et al. (2015).

3.5 Application of Non-Linear Compound Hawkes Models See Sect. 2.5

We also study other new Hawkes processes, namely, so-called non-linear general compound Hawkes processes (Swishchuk 2017), and apply them to model the price processes in the limit order books. We also prove Law of Large Numbers (LLN) and Functional Central Limit Theorems (FCLT) for these processes. The latter two FCLTs are applied to limit order books where we use these asymptotic methods to study the link between price volatility and order flow in our model by investigating the diffusion limits of these price processes. The volatilities of price changes are expressed in terms of parameters describing the arrival rates and price changes.

3.6 Application of Exponential General Compound Hawkes Models See Sect. 2.6

We consider option pricing with exponential general compound Hawkes model (EGCHM) such as analogues of Black-Scholes formula. Here, we present the analogue of Black-Scholes formula for European option price with EGCHM:

$$C = SN(d_+(T,S)) - Ke^{-rT}N(d_-(T,S)),$$

where

$$d_\pm(T,S) = \frac{1}{\sqrt{T\sigma^{*2}\frac{\lambda_\infty}{1-\alpha/\beta} + Ta^{*2}\frac{\lambda_\infty}{(1-\alpha/\beta)^3}}} \times$$
$$\left[\ln\frac{S}{K} + \left(r \pm \left(\frac{\sigma^{*2}}{2}\frac{\lambda_\infty}{1-\alpha/\beta} + \frac{na^{*2}}{2}\frac{\lambda_\infty}{(1-\alpha/\beta)^3}\right)\right)T\right],$$

where $N(x)$ is the normal distribution function, and all parameters in above formula are from initial model defined in II, F. See Guo et al. (2023) for more details.

3.7 Application of Multivariate Hawkes Model See Sect 2.7

We consider option pricing with exponential general compound Hawkes model (EGCHM) such as analogue of Margrabe's spread and basket option pricing formulas. Here we present Margrabe's spread option pricing formula (see Margrabe (1978).

We suppose that two risky assets $S_i(t)$ follow the following dynamics, respectively:

$$S_i(t) = S_i(0)\exp\{G_i(t)\},$$

where $G_i(t) := \sum_{k=1}^{N_i(t)} a_i(X_k^i)$ are two GCHP, $i = 1, 2$, $N_i(t)$ are two independent Hawkes processes with background intensities λ_i and self-exciting functions $\mu_i(t)$, X_k^i are two independent discrete-time finite or infinite state Markov chains with state space $X = \{1, 2, 3, ..., N\}$ or $X = \{1, 2, ..., N, ...\}$, respectively.

Margrabe's spread option valuation formula (see Margrabe (1978)) for two Hawkes-based assets $S_i(t)$ has the following form

$$V(0, S_1(0), S_2(0)) = S_1(0)N(d_1) - S_2(0)N(d_2),$$

where

$$d_{1,2} = \frac{\ln(S_1(0)/S_2(0)) \pm \frac{1}{2}\bar{\sigma}_{exp}^2}{\bar{\sigma}_{exp}},$$

and

$$\bar{\sigma}_{exp}^2 = Var(\bar{\sigma}_1\hat{W}_i(T) - \bar{\sigma}_2\hat{W}_2(T)) = T[\bar{\sigma}_1^2 + \bar{\sigma}_2^2 - 2\rho\bar{\sigma}_1\bar{\sigma}_2].$$

Here, ρ is the correlation between two assets, $\bar{\sigma}_i$ are volatility for given assets. The basket option pricing formula has the following form:

$$V_{\text{Basket}}(T) = e^{-rT}\theta\left(e^{\tilde{m}+\frac{1}{2}\tilde{v}^2}N(\theta d_1) - K^*N(\theta d_2)\right),$$

where $d_1 = \frac{\tilde{m}-\log K^*+\tilde{v}^2}{\tilde{v}}$, $d_2 = d_1 - \tilde{v}$, coefficients $\tilde{m}$, $\tilde{v}^2$ and K^* are defined by the parameters of the initial model, for example, $K^* = K - (\mathrm{E}(S_{basket}(T)) - \mathrm{E}(S_{geometric}(T)))$. Here: $S_{basket}(T))$ is the basket weighted price for several assets, and $S_{geometric}(T)$ is the geometric price for several assets, also $\theta = 1$ for call and $\theta = -1$ for put option.

4 Some Numerics

Which model is the best for a specific set of big data? If we take, e.g., 5 stocks EBAY, FB, MU, PCAR, SMH from Cartea et al. (2015), then the answer is (see Swishchuk and He (2019)):
EBAY: N-state GCHM ($N = 8$)
FB: 2-state GCHM ($N = 2$)
MU: 2-state GCHM ($N = 2$)
PCAR: N-state GCHM ($N = 15$)
SMH: N-state GCHM ($N = 9$)

Figure 1 shows the relationship between European call option price C (see formula for C in III, F), spot price S_t, and time t (all parameters were calibrated using LOBster data).

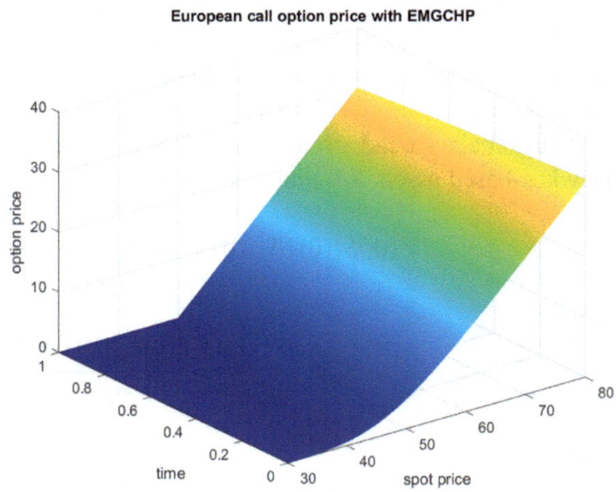

Fig. 1 European call option price

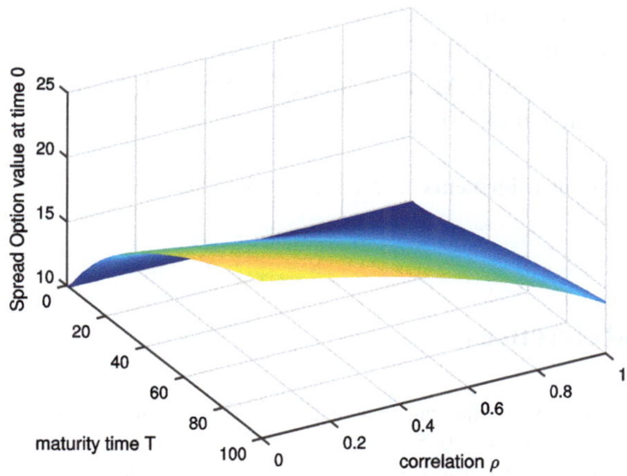

Fig. 2 Margrabe's spread option price

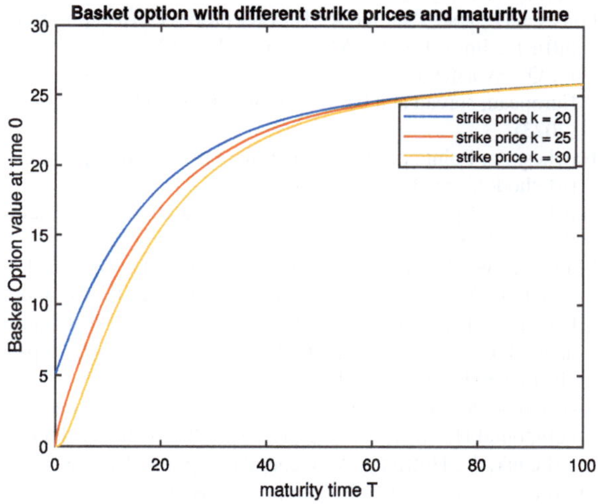

Fig. 3 Basket option prices for different strike prices

Figure 2 shows the Margrabe's spread option price (see III, G) (all parameters were calibrated using LOBster data).

Figure 3 shows the basket option price for three stocks with different strike prices (see III, G) (all parameters were calibrated using LOBster data).

Remark. The description of LOBster data may be found here: https://lobsterdata.com/info/DataSamples.php. The description of the LOBster data actual files ('message' and 'orderbook') may be found here: http://LOBSTER.wiwi.hu-berlin.de. The financial instruments traded on the Xetra and Frankfurt markets (Deutsche Boerse Group) on September 23, 2013 (15 assets) may be found here:

http://datashop.deutsche-boerse.com/1016/en. CISCO assets (November 3d, 2014) and others (EBAY: eBay Inc.; FB: Facebook Inc.; MU: MicronTechnology Inc. PCAR: PACCAR Inc.; SMH: VANECK Inc.) may be found in Cartea et al. (2015) (Cartea et al. 2015).

Acknowledgements The author thanks NSERC for continuing support.

References

Cartea, A., Jaimungal, S., Penalva, J.: Algorithmic and High-Frequency Trading. Cambridge University Press (2015)

Cont, R., de Larrard, A.: Price dynamics in a Markovian limit order market. SIAM J. Financial Math. **4**(1), 1–25 (2013)

Fonseca, D., José, Zaatour, R.: Hawkes process: fast calibration, application to trade clustering, and diffusive limit. J. Futur. Mark. **34**, 548–79 (2013)

Guo, Q., Swishchuk, A., Remillard, B.: Multivariate Hawkes-based Models in limit order book: European, spread and basket option pricing. Int. J. Theor. Appl. Financ. Dedicated to Tom Hurd (2023)

Hawkes, A.: Point spectra of some mutually-exciting point processes. J. R. Stat. Soc. Ser. B Stat. (Methodol.) **33**, 438–43 (1971)

Hawkes, A.: Spectra of some self-exciting and mutually exciting point processes. Biometrika **58**, 83–90 (1971)

Margrabe, W.: The value to exchange one asset for another. J. Financ. **33**, 177–186 (1978)

Swishchuk, A., Cassilas, J., Elliott, R., Remillard, B.: Compound Hawkes processes in limit order books. In: Handbook of Applied Econometrics: Financial Mathematics, Volatility and Covariance Modelling. Routledge Taylor and Francis (2019). https://papers.ssrn.com/sol3/papers.cfm?abstract SSRN id:2987943

Swishchuk, A., He, Q.: Quantitative and comparative analyses of limit order books with general compound Hawkes processes. Risks 5 (2019). https://doi.org/10.3390/risksxx010005

Swishchuk, A., Huffman, A.: General compound Hawkes processes in LOB. Risks (2020)

Swishchuk, A.: General Compound Hawkes Processes in Limit Order Books (2017). arXiv:1706.07459

Swishchuk, A.: Risk Model Based on Compound Hawkes Process. Wilmott (2018)

Swishchuk, A.: Risk processes based on general compound Hawkes processes. In: Paper presented at IME 2017 Congress, Vienna, Austria, pp. 3–5 (2017). https://fam.tuwien.ac.at/events/ime2017/IME2017_abstracts.pdf

Swishchuk, A.: Modelling of limit order books by general compound Hawkes processes with implementations. Methodol. Comput. Appl. Probab. **23**, 299–428 (2020)

Swishchuk, A., Vadori, N.: Semi-Markovian modeling of limit order markets. SIAM J. Financ. Math. **8**(1), 240–273 (2017)

Swishchuk, A., Hoffmeister, T., Sera, K., Schmidt, J.: General semi-Markov model for limit order books. Int. J. Theor. Appl. Financ. **20**, 1750019 (2017)

Incomplete Market Analysis of Optimal Consumption and Robust Portfolio for the 4/2 Stochastic Volatility Model

Yuyang Cheng and Marcos Escobar-Anel

Abstract In this paper, we study an expected utility portfolio optimization problem involving consumption and robustness analysis for an incomplete market investor. We employ the preferable proportion-to-volatility market price of risk for this analysis in the setting of a 4/2 stochastic volatility model. Due to the non-affine nature, the solution for the value function involves confluent hypergeometric functions similar to those needed for option pricing within the 4/2 model. The paper explores the cases where closed-form solutions are available, providing complementary numerical analyses.

Keywords 4/2 stochastic volatility model · CRRA utility and consumption · Optimal portfolio choice · Heston's model · 3/2 model · Robust portfolio analysis

1 Introduction

In the seminal work of Merton (1980) (Eqs. II.1–II.8), the excess return of a security, i.e., its risk premium, was recommended to be proportional to powers of the volatility. Three models were proposed, all presented in terms of the ratio of the excess return and the volatility of the security, i.e., the market price of risk (MPR). The first model, type I, assumed a MPR proportional to volatility (i.e., power $1/2$). This model implies that each risk factor earns a risk premium that is proportional to the variance of the factor's return. This is our choice of model for our paper.

The specification of MPRs plays a very important role in expected utility portfolio optimization. For example, Kraft (2005) solved the portfolio optimization problem

Y. Cheng
Canadian Imperial Bank of Commerce, Toronto, ON, Canada
e-mail: Yuyang.Cheng@cibc.com

M. Escobar-Anel (✉)
Department of Statistical and Actuarial Sciences, University of Western Ontario, London, ON, Canada
e-mail: marcos.escobar@uwo.ca

E. Braverman and A. Swishchuk (eds.), *Delay and Stochastic Differential Equations*, ICIAM2023 Springer Series 12, https://doi.org/10.1007/978-981-95-7310-3_6

for MPR of type I in a setting of CRRA (power) utility, in an incomplete market with finite horizon for the Heston model. i.e., the 1/2 model. Chacko and Viceira (2005) considered the optimal investment and consumption problem in an incomplete market for the 3/2 model of Heston (1997) with Epstein-Zin-Weil recursive utility, an infinite horizon and a MPR of type I, obtaining an approximated solution.

Our paper presents the very first closed-form analysis for type I MPR on the recently proposed 4/2 model (see Grasselli 2017) with a finite horizon, consumption and ambiguity-aversion, in a setting of incomplete markets. As a by-product, it provides the first analytical solution to type I MPR on the 3/2 model.

Several recent studies have been conducted on the 4/2 model for various MPRs. For instance, Cheng and Escobar-Anel (2021) explored the optimal investment problem for a risk-averse investor in both incomplete and complete markets in the absence of consumption. The authors employed the same MPR for the 1/2 and 3/2 components—proportional to $\sqrt{v_t}$, the driver of variance. The work of Cheng and Escobar-Anel (2023d) considered an investor that is not only risk-averse, but also ambiguity-averse, the work did not include consumption. Similarly, Cheng and Escobar-Anel (2023a) covered all the solvable, one dimensional, cases in the absence of consumption, while Cheng and Escobar-Anel (2023c) studied the problem with consumption but only for the complete market case.

Solving the optimal consumption and asset allocation with the advanced 4/2 model for a type I MPR is challenging. Our closed-form solutions is non-affine hence an important departure from the exiting literature and similar to the complete market case of Cheng and Escobar-Anel (2023c).

The paper is structured as follows. Section 2 describes the 4/2 model under consideration and the derivatives needed in the portfolio. Section 3 presents and solves the consumption and terminal wealth expected utility problem for a risk-averse investor. Section 4 then extends the problem to an ambiguity-averse investor for the incomplete market case (Sect. 4.1). Section 5 studies and implements the top three main cases numerically. In particular, Sect. 5.1 covers an ambiguity-averse investor in the presence of consumption for incomplete markets. Finally, Sect. 6 concludes this chapter. All proofs are provided with details in a complementary appendix.

2 Model Formulation

We assume the stochastic processes describing the financial market are defined on a complete probability space $(\Omega, \mathcal{F}, \mathbb{P})$ with a right-continuous filtration $\{\mathcal{F}_t\}_{t\in[0,T]}$. The price process S_t of the risky asset follows the so-called 4/2 model:

$$\frac{dS_t}{S_t} = \left(r + \rho\bar{\lambda}_1(a\sqrt{v_t} + \frac{b}{\sqrt{v_t}})^2 + \sqrt{1-\rho^2}\bar{\lambda}_2(a\sqrt{v_t} + \frac{b}{\sqrt{v_t}})^2\right) dt + (a\sqrt{v_t} + \frac{b}{\sqrt{v_t}})dW_t, \quad (1)$$

$$dv_t = \kappa\,(\theta - v_t)\,dt + \sigma\sqrt{v_t}dZ_{1t}, \; v(0) = v_0 > 0, \quad (2)$$

where v_t is the variance driver, which follows a CIR process with mean-reversion rate $\kappa > 0$, long-run mean $\theta > 0$, and volatility of volatility $\sigma > 0$. The Feller condition (i.e., $2\kappa\theta \geq \sigma^2$) is also imposed to keep the process v_t strictly positive. The standard Brownian motions (BM) W_t in the dynamics of the risky asset S_t and Z_{1t} in the dynamics of variance driver v_t are correlated with parameter $\rho \in (-1, 1)$. Thus, we will write $dW_t = \rho dZ_{1t} + \sqrt{1-\rho^2} dZ_{2t}$, where Z_{2t} is another standard BM, independent of Z_{1t}. The variance, denoted by $z_t = z(v_t)$ is given as follows:

$$z_t = \left(a\sqrt{v_t} + \frac{b}{\sqrt{v_t}} \right)^2 = a^2 v_t + \frac{b^2}{v_t} + 2ab. \tag{3}$$

This setting implies market prices of risk with the following representation:

$$\begin{cases} \lambda_1(v_t) = \bar{\lambda}_1 \left(a\sqrt{v_t} + \frac{b}{\sqrt{v_t}} \right), \\ \lambda_2(v_t) = \bar{\lambda}_2 \left(a\sqrt{v_t} + \frac{b}{\sqrt{v_t}} \right), \\ \lambda(v_t) = \rho\lambda_1(v_t) + \sqrt{1-\rho^2}\lambda_2(v_t) = \bar{\lambda} \left(a\sqrt{v_t} + \frac{b}{\sqrt{v_t}} \right), \end{cases} \tag{4}$$

where $\bar{\lambda} = \left(\rho\bar{\lambda}_1 + \sqrt{1-\rho^2}\bar{\lambda}_2 \right)$, a and b are positive constants, $\bar{\lambda}_1$ and $\bar{\lambda}_2$ are constant. $\lambda_1(v_t)$ is the market price of variance risk, and $\lambda(v_t)$ is the market price of stock risk. Moreover, $\lambda_2(v_t)$ can be interpreted as the market price of stock idiosyncratic risk (i.e., with respect to Z_{2t}). Note that in this form of market price of risk, the excess return of the risky asset is proportional to its variance as recommended in the economics literature; see Merton (1980) Eq. (II.6), type I.

As for the market price of variance risk $\lambda_1(v_t)$, we use Ito's lemma to create the process of the variance:

$$dz_t = (...)dt + \left(a - \frac{b}{v_t}\right)\left(a\sqrt{v_t} + \frac{b}{\sqrt{v_t}} \right) dZ_{1t}. \tag{5}$$

Hence our choice of market price of variance risk is: $\bar{\lambda}_1 \left(a\sqrt{v_t} + \frac{b}{\sqrt{v_t}} \right) = \bar{\lambda}_1 \sqrt{z_t}$. That is, it is proportional to the volatility of the asset. This is similar to the proposal in Heston (1993). Equations (1), and (2) are considered as the reference model.

3 Portfolio Optimization Under Expected Utility Theory (EUT)

We consider the investor exhibits CRRA utility for both intermediate consumption and terminal wealth with the same risk-aversion level γ. That is,

$$u(c) = \epsilon_1 \frac{c^\gamma}{\gamma}, \quad u(X_T) = \epsilon_2 \frac{X_T^\gamma}{\gamma}, \tag{6}$$

where coefficients ϵ_1 and ϵ_2 are non-negative. The ratio ϵ_1/ϵ_2 indicates the relative importance of intermediate consumption and terminal wealth, and it thus affects decision-making (optimal strategy). Without loss of generality, we can set $\epsilon_2 = 1$, and let ϵ_1 determine the relative importance ratio.

The objective of the investor is to maximize their utility from intermediate consumption c_t and terminal wealth X_T; therefore the reward functional for the investor is defined as follows:

$$w(x, v, t; \pi, c) = \mathbb{E}_{x,v,t}\left[\epsilon_1 \int_t^T e^{-\delta(\tau-t)} \frac{c_\tau^\gamma}{\gamma} d\tau + e^{-\delta(T-t)} \frac{X_T^\gamma}{\gamma}\right], \tag{7}$$

where δ is a discount rate, and the goal is

$$J(x, v, t) = \sup_{(\pi,c)\in\mathcal{U}} w(x, v, t; \pi, c), \tag{8}$$

where $J(x, v, t)$ is the value function and the space $\mathcal{U}$ of admissible controls $\{\pi_t, c_t\}_{t\in[0,T]}$ with $\pi_t \in \mathbb{R}$, $c_t \in \mathbb{R}^+$, is the set of feedback strategies that satisfy standard conditions (see Escobar et al. 2015).

3.1 *Incomplete Market Analysis*

Given the popularity of incomplete market situations, we describe the solution for this setting next. Let π_t be the fraction of wealth invested in the stock and $(1 - \pi_t)$ be the remaining portion of wealth invested in the money account, while c_t is the consumption at time t. The wealth X_t of the investor follows

$$\begin{aligned} dX_t = X_t\left[r + \pi_t\left(\rho\bar{\lambda}_1(a\sqrt{v_t} + \frac{b}{\sqrt{v_t}})^2 + \sqrt{1-\rho^2}\bar{\lambda}_2(a\sqrt{v_t} + \frac{b}{\sqrt{v_t}})^2\right)\right]dt - c_t dt \\ + \pi_t X_t(a\sqrt{v_t} + \frac{b}{\sqrt{v_t}})(\rho dZ_{1t} + \sqrt{1-\rho^2} dZ_{2t}). \end{aligned} \tag{9}$$

The value function $J(x, v, t)$ satisfies the HJB equation:

$$\sup_{\pi,c}\Bigg\{u(c) - \delta J + J_t + \Bigg(x\Bigg[r + \pi\Bigg[\rho\bar{\lambda}_1(a\sqrt{v_t} + \frac{b}{\sqrt{v_t}})^2 + \sqrt{1-\rho^2}\bar{\lambda}_2(a\sqrt{v_t} + \frac{b}{\sqrt{v_t}})^2\Bigg]\Bigg] - c\Bigg)J_x$$
$$+ \frac{1}{2}x^2\left(a\sqrt{v_t} + \frac{b}{\sqrt{v_t}}\right)^2 \pi^2 J_{xx}$$
$$+ \kappa(\theta - v_t)J_v + \frac{1}{2}\sigma^2 v_t J_{vv} + \sigma x \pi \rho (a v_t + b) J_{xv}\Bigg\} = 0, \tag{10}$$

with boundary condition $J(x, v, T) = \frac{x^\gamma}{\gamma}$. Here, we conjecture $J(x, v, t) = \frac{x^\gamma}{\gamma}\Big(h(t, v)\Big)^{1-\gamma}$, with $h(T, v) = 1$ for all v. This leads to a PDE for h as follows:

$$(1-\gamma)(\epsilon_1)^{-\frac{1}{\gamma-1}} + (1-\gamma)h_t + \left[r - \frac{\delta}{\gamma} - \frac{1}{2}\frac{(\rho\bar{\lambda}_1 + \sqrt{1-\rho^2}\bar{\lambda}_2)^2(a\sqrt{v} + \frac{b}{\sqrt{v}})^2}{(\gamma-1)}\right]\gamma h$$
$$+ \left[\kappa(\theta - v)(1-\gamma) - \frac{\sigma\rho(1-\gamma)(\rho\bar{\lambda}_1 + \sqrt{1-\rho^2}\bar{\lambda}_2)(av + b)}{(\gamma-1)}\gamma\right]h_v$$
$$+ \left[\frac{1}{2}\sigma^2 v(1-\gamma)\gamma - \frac{1}{2}\gamma\frac{\sigma^2\rho^2(1-\gamma)^2 v}{(\gamma-1)}\right]\frac{h_v^2}{h} + \left[\frac{1}{2}\sigma^2 v(1-\gamma)\right]h_{vv} = 0. \tag{11}$$

The coefficient of term $\frac{h_v^2}{h}$ in Eq. (11) cannot be made zero under any parametric setting; that is, $\frac{1}{2}\sigma^2 v(1-\gamma)\gamma - \frac{1}{2}\gamma\frac{\sigma^2\rho^2(1-\gamma)^2 v}{(\gamma-1)} = \frac{1}{2}\sigma^2 v(1-\gamma)\gamma\Big(1+\rho^2\Big) \neq 0$. This means there is no closed-form solution in the presence of consumption. Nevertheless, in the absence of consumption, we can find a solution; the result is provided next.

We conjecture our value function follows

$$J(x, v, t) = \frac{x^\gamma}{\gamma}h(t, v), \tag{12}$$

where $h(T, v) = 1$ for all v, which leads to the PDE

$$h_t + \underbrace{\left[r\gamma - \frac{\gamma}{2}\frac{\Big(\rho\bar{\lambda}_1(a\sqrt{v_t} + \frac{b}{\sqrt{v_t}}) + \sqrt{1-\rho^2}\bar{\lambda}_2(a\sqrt{v} + \frac{b}{\sqrt{v}})\Big)^2}{(\gamma-1)}\right]}_{\triangleq V(v,t)}h + \left[-\frac{\gamma}{2}\frac{\sigma^2\rho^2 v}{(\gamma-1)}\right]\frac{h_v^2}{h}$$
$$+ \underbrace{\left[\kappa(\theta - v) - \gamma\frac{\Big(\rho\bar{\lambda}_1(a\sqrt{v_t} + \frac{b}{\sqrt{v_t}}) + \sqrt{1-\rho^2}\bar{\lambda}_2(a\sqrt{v} + \frac{b}{\sqrt{v}})\Big)}{(\gamma-1)}\sigma\rho\sqrt{v}\right]}_{\triangleq\Gamma(v,t) = \tilde{\kappa}\tilde{\theta} - \tilde{\kappa}v}h_v + \frac{1}{2}\sigma^2 v h_{vv} = 0, \tag{13}$$

where

$$
\begin{aligned}
\tilde{\kappa} &= \kappa + \frac{\gamma}{\gamma-1}\left(\rho\bar{\lambda}_1 + \sqrt{1-\rho^2}\bar{\lambda}_2\right)\sigma\rho a,\\
\tilde{\kappa}\tilde{\theta} &= \kappa\theta - \frac{\gamma}{\gamma-1}\left(\rho\bar{\lambda}_1 + \sqrt{1-\rho^2}\bar{\lambda}_2\right)\sigma\rho b.
\end{aligned} \tag{14}
$$

Proposition 3.1 (4/2 model in incomplete market, no consumption) *Let* $\epsilon_1 = 0$ *and*

$$
\begin{aligned}
\mu &= \frac{1}{2}\frac{\gamma}{\gamma-1}(\rho\bar{\lambda}_1 + \sqrt{1-\rho^2}\bar{\lambda}_2)^2 a^2,\\
\nu &= \frac{1}{2}\frac{\gamma}{\gamma-1}(\rho\bar{\lambda}_1 + \sqrt{1-\rho^2}\bar{\lambda}_2)^2 b^2.
\end{aligned} \tag{15}
$$

Assume the following four conditions

$$
\rho = 0,\ \ \tilde{\kappa}\tilde{\theta} \geq \frac{\sigma^2}{2},\ \ \mu > \frac{-\tilde{\kappa}^2}{2\sigma^2},\ \ \nu \geq -\frac{(2\tilde{\kappa}\tilde{\theta}-\sigma^2)^2}{8\sigma^2}. \tag{16}
$$

Then, the solution of the HJB Equation (10) *has representation* (12) *with*

$$
h(v,t) = \exp\left\{\left(r\gamma - \frac{\gamma}{\gamma-1}(\bar{\lambda}_1\rho + \bar{\lambda}_2\sqrt{1-\rho^2})^2 ab\right)(T-t)\right\} \times q(\tau, v; \alpha, \lambda, \mu, \nu), \tag{17}
$$

with $q(\tau, v; \alpha, \lambda, \mu, \nu)$ *as follows*

$$
\begin{aligned}
q(\tau, v; \mu, \nu) = {} & \left(\frac{\beta(\tau,v)}{2}\right)^{m+1} v^{-\frac{\tilde{\kappa}\tilde{\theta}}{\sigma^2}} K(\tau)^{-\left(\frac{1}{2}+\frac{m}{2}+\frac{\tilde{\kappa}\tilde{\theta}}{\sigma^2}\right)} \times e^{\frac{1}{\sigma^2}\left(\tilde{\kappa}^2\tilde{\theta}\tau - \sqrt{D}v\coth\left(\frac{\sqrt{D}\tau}{2}+\tilde{\kappa}v\right)\right)}\\
& \frac{\Gamma\left(\frac{1}{2}+\frac{m}{2}+\frac{\tilde{\kappa}\tilde{\theta}}{\sigma^2}\right)}{\Gamma(m+1)} \times {}_1F_1\left(\frac{1}{2}+\frac{m}{2}+\frac{\tilde{\kappa}\tilde{\theta}}{\sigma^2}, m+1, \frac{\beta(\tau,v)^2}{4K(\tau))}\right),
\end{aligned} \tag{18}
$$

with

$$
\begin{aligned}
m &= \frac{1}{\sigma^2}\sqrt{(2\tilde{\kappa}\tilde{\theta}-\sigma^2)^2 + 8\sigma^2\nu},\ D = \tilde{\kappa}^2 + 2\mu\sigma^2,\\
\beta(\tau, v) &= \frac{2\sqrt{Dv}}{\sigma^2\sinh\left(\frac{\sqrt{D}\tau}{2}\right)},\ K(\tau) = \frac{1}{\sigma^2}\left(\sqrt{D}\coth\left(\frac{\sqrt{D}\tau}{2}\right) + \tilde{\kappa}\right).
\end{aligned} \tag{19}
$$

Moreover, the optimal wealth exposure is given by

$$
(\pi)^\star = \frac{\bar{\lambda}_2}{1-\gamma}. \tag{20}
$$

See the proof in Appendix 1.

The previous result highlights limitations stemming from an extra, necessary condition (i.e., no leverage effect $\rho = 0$) to achieve a closed-form solution, which coincides with the celebrated Merton's solution.

4 Robust Consumption Portfolio Optimization Under EUT

Assume the investor is uncertain about the probability distribution for the reference model and consider a set of plausible, alternative models when making investment decisions. Specifically, the investor is uncertain about the distribution of Z_{1t} and Z_{2t}.

Let $e := (e_t^v, e_t^S)$ be an $\mathbb{R}^2$-valued $\mathcal{F}$-progressively measurable process and define the Radon-Nikodym derivative process by

$$\xi_t^e = \frac{d\mathbb{P}^e}{d\mathbb{P}}|\mathcal{F}_t = \exp\left\{ - \int_0^t \left(\frac{(e_\tau^v)^2 + (e_\tau^S)^2}{2} d\tau + e_\tau^v dZ_{1\tau} + e_\tau^S dZ_{2\tau} \right) \right\}. \quad (21)$$

According to Girsanov's theorem, the process

$$\begin{bmatrix} \widetilde{Z}_{1t} \\ \widetilde{Z}_{2t} \end{bmatrix} = \begin{bmatrix} \int_0^t e_\tau^v d\tau \\ \int_0^t e_\tau^S d\tau \end{bmatrix} + \begin{bmatrix} Z_{1t} \\ Z_{2t} \end{bmatrix}, \quad (22)$$

is a Wiener process under probability measure $\mathbb{P}^e$. Let $\varepsilon[0, T]$ denote the set of all $\mathcal{F}_t$-progressively measurable processes such that the process (21) is a well-defined Radon-Nikodym derivative process. This formulation of incorporating the investor's model uncertainty is actually allowing uncertainty on the drift of diffusion risk factors of the stock and its variance's driver (i.e., Z_{2t} and Z_{1t}, respectively).

The alternative model then follows:

$$\begin{aligned} \frac{dS_t}{S_t} &= \Big[r + \rho\bar{\lambda}_1 (a\sqrt{v_t} + \frac{b}{\sqrt{v_t}})^2 + \sqrt{1-\rho^2}\bar{\lambda}_2 (a\sqrt{v_t} + \frac{b}{\sqrt{v_t}})^2 - \rho(a\sqrt{v_t} + \frac{b}{\sqrt{v_t}}) e_t^v \\ &\quad - \sqrt{1-\rho^2}(a\sqrt{v_t} + \frac{b}{\sqrt{v_t}}) e_t^S \Big] dt + (a\sqrt{v_t} + \frac{b}{\sqrt{v_t}}) \left(\rho d\widetilde{Z}_{1t} + \sqrt{1-\rho^2} d\widetilde{Z}_{2t} \right), \\ dv_t &= \left(\kappa(\theta - v_t) - \sigma\sqrt{v_t} e_t^v \right) dt + \sigma\sqrt{v_t} d\widetilde{Z}_{1t}, \end{aligned} \quad (23)$$

The reward functional can be defined as in the previous section for a given probability measure $\mathbb{P}^e$:

$$w^e(x, v, t; \pi, c) = \mathbb{E}_{x,v,t}^{\mathbb{P}^e} \left[\epsilon_1 \int_t^T e^{-\delta(\tau - t)} \frac{c_\tau^\gamma}{\gamma} d\tau + e^{-\delta(T-t)} \frac{X_T^\gamma}{\gamma} \right]. \quad (24)$$

In the presence of a preference for robustness, the investor's objective is to minimize the penalty term and maximize his utility from intermediate consumption c_t and terminal wealth X_T:

$$J(x,v,t) = \sup_{\pi,c\in\mathcal{U}} \inf_{e^S,e^v\in\mathcal{V}} \left(w^e(x,v,t;\pi,c) + \mathbb{E}_t^{\mathbb{P}^e}\left[\int_t^T \left(\frac{(e_\tau^v)^2}{2\Phi_1(x,v,\tau)} + \frac{(e_\tau^S)^2}{2\Phi_2(x,v,\tau)} \right) d\tau \right] \right), \tag{25}$$

where $J(x, v, t)$ is the value function, and the last two terms serves as the penalty for deviating too far from the reference model. The space $\mathcal{V}$ of $\mathcal{F}_t$-adapted processes $\mathbf{e}_t = \{e^S, e^v\} \in \mathbb{R}^2$, is the set of perturbations; the space $\mathcal{U}$ of admissible controls $\{\pi_t, c_t\}_{t\in[0,T]}$ (i.e., $\pi_t \in \mathbb{R}$, $c_t \in \mathbb{R}^+$) is the set of feedback admissible strategies. The perturbations e_t^v and e_t^S in the penalty term are scaled by Φ_1 and Φ_2, respectively. That is, the larger the values of Φ_1 and Φ_2, the smaller the penalties for deviating from the reference model, which implies that the investor is more uncertain about the model. Following Maenhout (2004), we assume

$$\Phi_i = \frac{\phi_i}{\gamma \bar{J}(x,v,t)}, \quad i = 1, 2, \tag{26}$$

where $\phi_i > 0$ denotes the ambiguity-aversion parameters. In this specification, the optimal strategy would be independent of the current wealth level for a power utility investor, namely homothetic robustness in Maenhout (2004). Further, ϕ_1 can be interpreted as ambiguity aversion regarding the volatility driver, while ϕ_2 represents ambiguity about the stock process.

4.1 Incomplete Market

To provide a full analysis, and given the popularity of incomplete market solutions, here we address closed-form expressions in an incomplete market. Let π_t be the fraction of wealth invested in the stock and $(1 - \pi_t)$ be the remaining portion of wealth invested in the money account, while c_t is the consumption at time t. The wealth X_t of the investor follows

$$\begin{aligned} dX_t = X_t\Bigg[r + \pi_t \Bigg(\rho\bar{\lambda}_1 (a\sqrt{v_t} + \frac{b}{\sqrt{v_t}})^2 + \sqrt{1-\rho^2}\bar{\lambda}_2 (a\sqrt{v_t} + \frac{b}{\sqrt{v_t}})^2 - \rho(a\sqrt{v_t} + \frac{b}{\sqrt{v_t}}) e_t^v \\ - \sqrt{1-\rho^2}(a\sqrt{v_t} + \frac{b}{\sqrt{v_t}}) e_t^S \Bigg) \Bigg] dt - c_t dt + \pi_t X_t (a\sqrt{v_t} + \frac{b}{\sqrt{v_t}})(\rho d\widetilde{Z}_{1t} + \sqrt{1-\rho^2} d\widetilde{Z}_{2t}). \end{aligned} \tag{27}$$

The value function $J(x, v, t)$ satisfies the HJBI equation:

$$\begin{aligned} \sup_{\pi,c} \inf_{e^S,e^v} \Bigg\{ u(c) - \delta J + J_t + \Bigg(x\Bigg[r + \pi \Bigg[\rho\bar{\lambda}_1 (a\sqrt{v_t} + \frac{b}{\sqrt{v_t}})^2 + \sqrt{1-\rho^2}\bar{\lambda}_2 (a\sqrt{v_t} + \frac{b}{\sqrt{v_t}})^2 - \rho(a\sqrt{v_t} + \frac{b}{\sqrt{v_t}}) e_t^v \\ - \sqrt{1-\rho^2}(a\sqrt{v_t} + \frac{b}{\sqrt{v_t}}) e_t^S \Bigg]\Bigg] - c \Bigg) J_x + \frac{1}{2} x^2 \left(a\sqrt{v_t} + \frac{b}{\sqrt{v_t}} \right)^2 \pi^2 J_{xx} + \left[\kappa(\theta - v_t) - \sigma\sqrt{v_t} e_t^v \right] J_v \\ + \frac{1}{2}\sigma^2 v_t J_{vv} + \sigma x \pi \rho (a v_t + b) J_{xv} + \frac{(e^v)^2}{2\Phi_1} + \frac{(e^S)^2}{2\Phi_2} \Bigg\} = 0, \end{aligned} \tag{28}$$

with boundary condition $J(x, v, T) = \frac{x^\gamma}{\gamma}$ and $\Phi_i = \frac{\phi_i}{\gamma J}$, $i = 1, 2$. Similarly to the previous results, we conjecture our value function as follows:

$$J(x, v, t) = \frac{x^\gamma}{\gamma}\Big(h(t, v)\Big)^{1-\gamma}, \tag{29}$$

where $h(T, v) = 1$ for all v. This conjecture leads to the following PDE for h:

$$\begin{aligned}
&(1-\gamma)(\epsilon_1)^{-\frac{1}{\gamma-1}} + (1-\gamma)h_t + \left[r - \frac{\delta}{\gamma} + \frac{1}{2}\frac{(\rho\bar{\lambda}_1 + \sqrt{1-\rho^2}\bar{\lambda}_2)^2(a\sqrt{v} + \frac{b}{\sqrt{v}})^2}{\Big(\phi_1\rho^2 + \phi_2(1-\rho^2) - (\gamma-1)\Big)}\right]\gamma h \\
&+ \left[\kappa(\theta - v)(1-\gamma) + \frac{\sigma\rho(1-\gamma)(\rho\bar{\lambda}_1 + \sqrt{1-\rho^2}\bar{\lambda}_2)(av+b)}{\Big(\phi_1\rho^2 + \phi_2(1-\rho^2) - (\gamma-1)\Big)}(\gamma + \phi_1)\right]h_v \\
&+ \left[\frac{1}{2}\sigma^2 v(1-\gamma)(\gamma - \phi_1(1-\gamma)) + \frac{1}{2}(\gamma - \phi_1)\frac{\sigma^2\rho^2(1-\frac{\phi_1}{\gamma})(1-\gamma)^2 v}{\Big(\phi_1\rho^2 + \phi_2(1-\rho^2) - (\gamma-1)\Big)}\right]\frac{h_v^2}{h} \\
&+ \left[\frac{1}{2}\sigma^2 v(1-\gamma)\right]h_{vv} = 0.
\end{aligned} \tag{30}$$

See details in Appendix 2.

To produce solvable cases, we eliminate the term $\frac{h_v^2}{h}$, dividing both sides simultaneously by $(1-\gamma)$ leads to

$$\begin{aligned}
&(\epsilon_1)^{-\frac{1}{\gamma-1}} + h_t + \underbrace{\left[r - \frac{\delta}{\gamma} + \frac{1}{2}\frac{(\rho\bar{\lambda}_1 + \sqrt{1-\rho^2}\bar{\lambda}_2)^2(a\sqrt{v} + \frac{b}{\sqrt{v}})^2}{\Big(\phi_1\rho^2 + \phi_2(1-\rho^2) - (\gamma-1)\Big)}\right]\frac{\gamma}{1-\gamma}}_{\triangleq V(v,t)} h \\
&+ \underbrace{\left[\kappa(\theta - v) + \frac{\sigma\rho(\rho\bar{\lambda}_1 + \sqrt{1-\rho^2}\bar{\lambda}_2)(av+b)}{\Big(\phi_1\rho^2 + \phi_2(1-\rho^2) - (\gamma-1)\Big)}(\gamma + \phi_1)\right]}_{\triangleq \Gamma(v,t) = \tilde{\kappa}\tilde{\theta} - \tilde{\kappa}v} h_v + \frac{1}{2}\sigma^2 v h_{vv} = 0,
\end{aligned} \tag{31}$$

where

$$\begin{aligned}
\tilde{\kappa}\tilde{\theta} &= \frac{\sigma\rho(\rho\bar{\lambda}_1 + \sqrt{1-\rho^2}\bar{\lambda}_2)b}{\Big(\phi_1\rho^2 + \phi_2(1-\rho^2) - (\gamma-1)\Big)}(\gamma + \phi_1) + \kappa\theta, \\
\tilde{\kappa} &= -\frac{\sigma\rho(\rho\bar{\lambda}_1 + \sqrt{1-\rho^2}\bar{\lambda}_2)a}{\Big(\phi_1\rho^2 + \phi_2(1-\rho^2) - (\gamma-1)\Big)}(\gamma + \phi_1) + \kappa.
\end{aligned} \tag{32}$$

The main result is given next.

Proposition 4.1 (4/2 model in incomplete market, robustness) *Let*

$$\begin{aligned}\mu &= \frac{1}{2}\frac{\gamma}{1-\gamma}\frac{(\bar{\lambda}_1\rho + \bar{\lambda}_2\sqrt{1-\rho^2})^2 a^2}{(\gamma-1) - \phi_1\rho^2 - \phi_2(1-\rho^2)},\\ \nu &= \frac{1}{2}\frac{\gamma}{1-\gamma}\frac{(\bar{\lambda}_1\rho + \bar{\lambda}_2\sqrt{1-\rho^2})^2 b^2}{(\gamma-1) - \phi_1\rho^2 - \phi_2(1-\rho^2)}.\end{aligned} \tag{33}$$

Assume

$$\gamma - \phi_1(1-\gamma) = \frac{-\rho^2(\gamma-\phi_1)^2(1-\gamma)}{\gamma\Big(\phi_1\rho^2 + \phi_2(1-\rho^2) - (\gamma-1)\Big)}, \tag{34}$$

and $\tilde{\kappa}$, $\tilde{\theta}$, μ, *and* ν *in Equation* (33) *satisfy conditions* (35).

$$\tilde{\kappa}\tilde{\theta} \geq \frac{\sigma^2}{2}, \quad \mu > \frac{-\tilde{\kappa}^2}{2\sigma^2}, \quad \nu \geq -\frac{(2\tilde{\kappa}\tilde{\theta} - \sigma^2)^2}{8\sigma^2}, \tag{35}$$

Then, a candidate solution of the HJBI equation (28) *is* $J(x, v, t) = \frac{x^\gamma}{\gamma}\Big(h(t, v)\Big)^{1-\gamma}$, *where* h *admits the representation:*

$$h(v, t) = (\epsilon_1)^{-\frac{1}{\gamma-1}} \int_t^T g(v_{\tilde{\tau}}, \tilde{\tau})d\tilde{\tau} + g(v, \tau), \tag{36}$$

$$g(v, \tau) = \exp\left\{\frac{\gamma}{1-\gamma}\left(r - \frac{\delta}{\gamma} - \frac{\bar{\lambda}_1^2 ab}{\gamma-1-\phi_1} - \frac{\bar{\lambda}_2^2 ab}{\gamma-1-\phi_2}\right)(T-t)\right\} \times q(\tau, v; \alpha, \lambda, \mu, \nu), \tag{37}$$

with $\tau(t) = T - t$, *and* $q(\tau, v; \alpha, \lambda, \mu, \nu)$ *follows from Equation* (18) *with associated* m, D, β, *and* K. *Moreover, the optimal consumption-wealth ratio and stock allocation are given by*

$$\begin{aligned}\left(\frac{c}{x}\right)^\star &= h^{-1}(\epsilon_1)^{-\frac{1}{\gamma-1}},\\ (\pi)^\star &= \frac{\Big(\rho\bar{\lambda}_1 + \sqrt{1-\rho^2}\bar{\lambda}_2\Big)}{\Big(\phi_1\rho^2 + \phi_2(1-\rho^2) - (\gamma-1)\Big)}\\ &\quad + \frac{(1-\frac{\phi_1}{\gamma})\sigma\rho\sqrt{v}(1-\gamma)h_v}{\Big(\phi_1\rho^2 + \phi_2(1-\rho^2) - (\gamma-1)\Big)\Big(a\sqrt{v} + \frac{b}{\sqrt{v}}\Big)h}.\end{aligned} \tag{38}$$

The worst case measure is determined by

$$
\begin{aligned}
(e^v)^\star &= \phi_1\big[(\pi)^\star\rho\big(a\sqrt{v}+\frac{b}{\sqrt{v}}\big)+\sigma\sqrt{v}\frac{1-\gamma}{\gamma}\frac{h_v}{h}\big],\\
(e^S)^\star &= \phi_2\big[(\pi)^\star\sqrt{1-\rho^2}\big(a\sqrt{v}+\frac{b}{\sqrt{v}}\big)\big].
\end{aligned}
\tag{39}
$$

See the proof in Appendix 2.

Interestingly, here we manage to produce a closed-form solution in the presence of consumption, which was not possible in the non-robust analysis of Sect. 3.1. Moreover, the robust incomplete market solution is not as simple as the non-robust Merton-like solution in Proposition 3.1—there is a non-myopic term that does not go away as ϕ goes to zero. This is an indication that in the presence of consumption, the non-robust solution shall also have had a non-myopic term.

In an incomplete market, in the absence of consumption, the solution is also simpler. Nonetheless we need to impose the condition[1]:

$$
\phi_1 = \frac{-(\gamma-\phi_1)^2\rho^2}{(\gamma-1)-\phi_1\rho^2-\phi_2(1-\rho^2)}. \tag{40}
$$

Then, $J(x,v,t)=\frac{x^\gamma}{\gamma}h(t,v)$ where h solves the PDE

$$
\begin{aligned}
0 = h_t &+ \underbrace{\left[r\gamma - \frac{\gamma}{2}\frac{\Big(\rho\bar{\lambda}_1(a\sqrt{v}+\frac{b}{\sqrt{v}})+\sqrt{1-\rho^2}\bar{\lambda}_2(a\sqrt{v}+\frac{b}{\sqrt{v}})\Big)^2}{(\gamma-1)-\phi_1\rho^2-\phi_2(1-\rho^2)}\right]}_{\triangleq V(v,t)} h\\
&+ \underbrace{\left[\kappa(\theta-v)-(\gamma-\phi_1)\frac{\Big(\rho\bar{\lambda}_1(a\sqrt{v}+\frac{b}{\sqrt{v}})+\sqrt{1-\rho^2}\bar{\lambda}_2(a\sqrt{v}+\frac{b}{\sqrt{v}})\Big)}{(\gamma-1)-\phi_1\rho^2-\phi_2(1-\rho^2)}\sigma\rho\sqrt{v}\right]}_{\triangleq \Gamma(v,t)=\tilde{\kappa}\tilde{\theta}-\tilde{\kappa}v} h_v\\
&+ \frac{1}{2}\sigma^2 v h_{vv},
\end{aligned}
\tag{41}
$$

where

$$
\begin{aligned}
\tilde{\kappa}\tilde{\theta} &= \kappa\theta-(\gamma-\phi_1)\frac{\Big(\rho\bar{\lambda}_1+\sqrt{1-\rho^2}\bar{\lambda}_2\Big)}{(\gamma-1)-\phi_1\rho^2-\phi_2(1-\rho^2)}\sigma\rho b,\\
\tilde{\kappa} &= \kappa+(\gamma-\phi_1)\frac{\Big(\rho\bar{\lambda}_1+\sqrt{1-\rho^2}\bar{\lambda}_2\Big)}{(\gamma-1)-\phi_1\rho^2-\phi_2(1-\rho^2)}\sigma\rho a.
\end{aligned}
\tag{42}
$$

[1] The condition leads to a feasible region of values for ϕ_1 and ϕ_2 given ρ.

If $\tilde{\kappa}$, $\tilde{\theta}$, μ, and ν satisfy conditions (35), then h admits a Feynman-Kac representation and can be solved as follows:

$$h(v,t) = \exp\left\{ \left(r\gamma - \gamma \frac{(\bar{\lambda}_1\rho + \bar{\lambda}_2\sqrt{1-\rho^2})^2 ab}{(\gamma-1) - \phi_1\rho^2 - \phi_2(1-\rho^2)} \right)(T-t) \right\} \times q(\tau, v; \alpha, \lambda, \mu, \nu), \tag{43}$$

with parameters

$$\begin{aligned} \mu &= \frac{1}{2}\frac{(\bar{\lambda}_1\rho + \bar{\lambda}_2\sqrt{1-\rho^2})^2 a^2\gamma}{(\gamma-1) - \phi_1\rho^2 - \phi_2(1-\rho^2)}, \\ \nu &= \frac{1}{2}\frac{(\bar{\lambda}_1\rho + \bar{\lambda}_2\sqrt{1-\rho^2})^2 b^2\gamma}{(\gamma-1) - \phi_1\rho^2 - \phi_2(1-\rho^2)}, \end{aligned} \tag{44}$$

where $q(T-t, v; \alpha, \lambda, \mu, \nu)$ comes from Eq. (18) with associated m, D, β, and K. The optimal wealth exposures are given by

$$(\pi)^\star = \frac{\sqrt{v}\left(\frac{\phi_1-\gamma}{\gamma}\sigma\rho\right)h_v - \left(\rho\bar{\lambda}_1(a\sqrt{v_t} + \frac{b}{\sqrt{v_t}}) + \sqrt{1-\rho^2}\bar{\lambda}_2(a\sqrt{v} + \frac{b}{\sqrt{v}})\right)h}{(a\sqrt{v} + \frac{b}{\sqrt{v}})\left[(\gamma-1) - \phi_1\rho^2 - \phi_2(1-\rho^2)\right]h}. \tag{45}$$

The worst case measure is determined by

$$\begin{aligned} (e^v)^\star &= \phi_1\big[(\pi)^\star\rho\big(a\sqrt{v} + \frac{b}{\sqrt{v}}\big) + \sigma\sqrt{v}\frac{1-\gamma}{\gamma}\frac{h_v}{h}\big], \\ (e^S)^\star &= \phi_2\big[(\pi)^\star\sqrt{1-\rho^2}\big(a\sqrt{v} + \frac{b}{\sqrt{v}}\big)\big]. \end{aligned} \tag{46}$$

5 Numerical Analysis

Here we present an incomplete market case with ambiguity-aversion and consumption.

Note that, we cannot use the estimation results of the "drift group" from Cheng and Escobar-Anel (2021) because of the new choice of MPR for 4/2 and 3/2 models. To accommodate to our choice of MPR, we re-estimate the rate of market price of risk $\bar{\lambda}$ for each model by fixing the excess return at $v_t = \theta$ (long-term value), in line with Cheng and Escobar-Anel (2021). Then, we follow the procedure of Cheng and Escobar-Anel (2021) and substitute $\bar{\lambda}$ into the regression to update ρ for each model. The estimation results and the other baseline parameters are presented in Tables 1 and 2, respectively. In this section, we set $\bar{\lambda}_2 = 2$, and solve for $\bar{\lambda}_1$ for each model according to the relationship $\bar{\lambda} = \rho\bar{\lambda}_1 + \sqrt{1-\rho^2}\,\bar{\lambda}_2$.

Table 1 Estimates among the various models

	4/2 Model	3/2 Model	Heston
$\hat{\kappa}$	7.3479	6.9884	14.6290
$\hat{\theta}$	0.0328	0.0323	0.0315
$\hat{\sigma}$	0.6612	0.3760	0.5210
$\hat{a}$	0.9051	0	1
$\hat{b}$	0.0023	0.0268	0
ρ	–0.7689	0.7910	–0.8129
$\bar{\lambda}$	3.0176	4.2973	2.8689
Theoretical leverage ($v_t = \theta$)	–0.7689	–0.7910	–0.8129

Table 2 Baseline parameters

r	δ	γ	v_0	t	T	ϵ	v_N
0.05	0.02	–0.5	0.04	0	10	1	0.04

5.1 Incomplete Market Analysis with Consumption for Ambiguity-Averse Investors

This last analysis also requires a parametric constraint on the levels of ambiguity-aversion and risk-aversion; see Eq. (34). The constraint is

$$\gamma - \phi_1(1-\gamma) = \frac{-\rho^2(\gamma-\phi_1)^2(1-\gamma)}{\gamma\Big(\phi_1\rho^2 + \phi_2(1-\rho^2) - (\gamma-1)\Big)}. \tag{47}$$

To fulfill this condition, we set the risk-aversion level $\gamma = 0.5$, and the ambiguity level $\phi_1 = 3$, and then solve for ϕ_2 $(= 3.5)$ accordingly.

The plots of the optimal consumption-wealth ratio c/x and the optimal allocation π versus (a) SD and (b) investment horizon T are presented in Figs. 1, 2, 3 and 4, respectively. Generally speaking, Figs. 1 and 2 show that the investor who follows the 3/2 model benefits the most when allowed to hedge the variance and thus complete the market. Specifically, in a complete market, the 3/2 model leads to more consumption while in an incomplete market, the 3/2 model leads to the least consumption among models.

As Liu (2010), Maenhout (2006), and Jiangyan and Zhang (2021) pointed out, a robust investor with risk-aversion level γ and ambiguity-aversion level ϕ will follow the same strategy as that of a non-robust investor with risk-aversion level $\gamma + \phi$. Hence, the conservative behavior of the 3/2 model may be explained by its sensitivity to the risk-aversion level. In comparison, the Heston and 4/2 models provide a relatively robust and conservative consumption rate with 4/2 becoming

Fig. 1 c/x versus SD

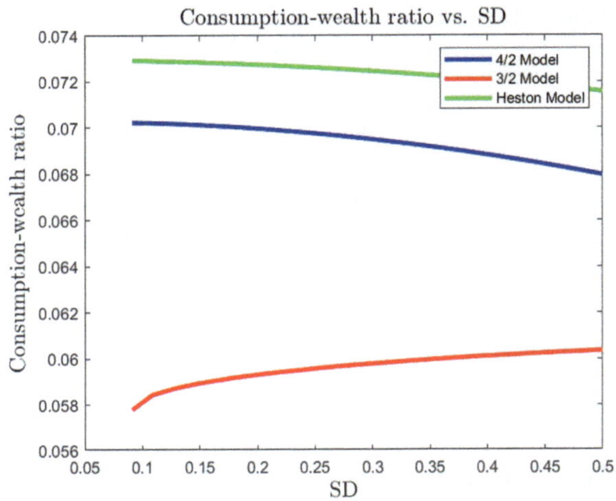

Fig. 2 c/x versus T

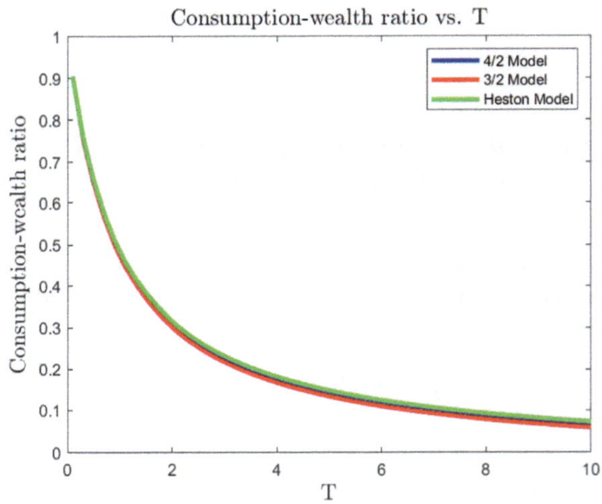

closer to the Heston model in highly volatile markets. Figures 3 and 4 illustrate that the 4/2 model recommends a strategy that lies between the Heston and 3/2 models by being more aggressive than the Heston model but not as aggressive as the 3/2 model.

The sensitivity analyses of the optimal consumption-wealth ratio c/x and optimal allocation π to the parameters a and b are explored in Figs. 5, 6, 7, and 8 respectively. From Figs. 5 and 6, it is interesting to find that b changing in $[0, 0.01]$ can affect the consumption-wealth ratio as much as that of a changing in $[0, 2]$, thus confirming the larger impact of b on consumption. Furthermore, the risky asset allocation decreases in both a and b with higher sensitivity to a.

Fig. 3 π versus SD

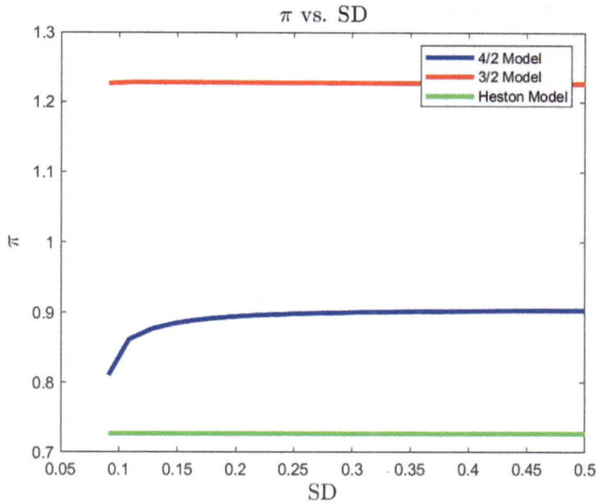

Fig. 4 π versus T

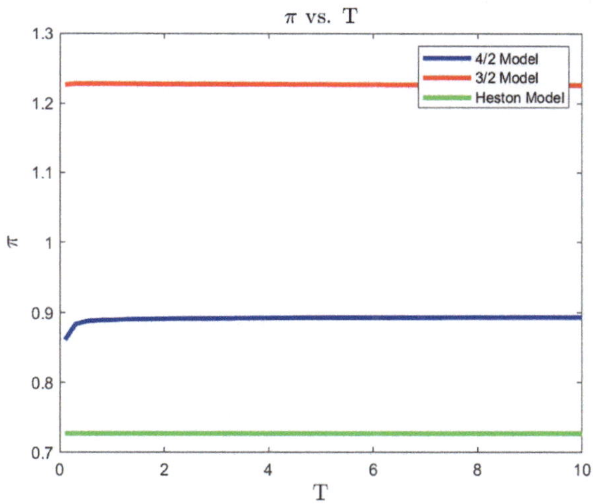

6 Conclusion

In this paper, an optimal investment problem for a risk-averse investor under the 4/2 SV model and a 4/2-structured MPR is considered by combining terminal wealth utility maximization with utility on consumption, incompleteness in the market and ambiguity-aversion. Although the non-affine nature of the 4/2 volatility and the 4/2-structured MPR is challenging, we found closed-form solutions under certain conditions. In the numerical part, we present and compare the portfolio strategies

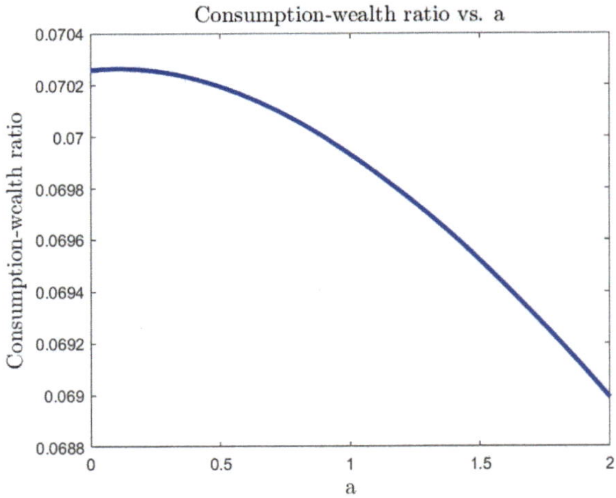

Fig. 5 $\frac{c}{x}$ versus a

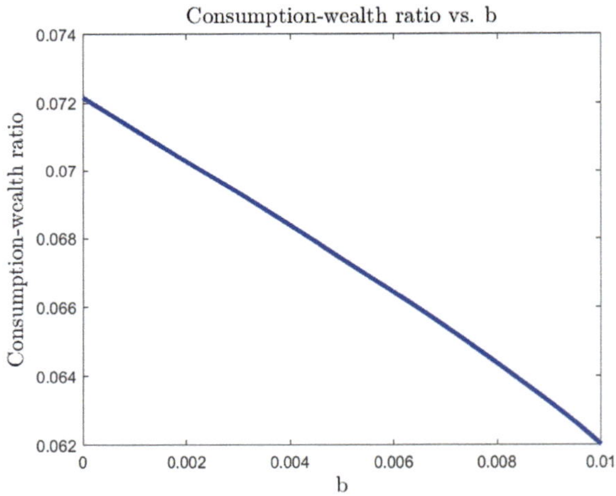

Fig. 6 $\frac{c}{x}$ versus b

recommended by the 4/2, 3/2, and 1/2 models for an investor who cares about both consumption and ambiguity in an incomplete market using real-data parameters. The 4/2 and 1/2 models generally behave similarly in wealth exposures compared to that of 3/2 model. The 4/2 model behaves like an average by lying in-between the Heston and 3/2 models in consumption and investment recommendation in an incomplete market. The work can be extended to multidimensions along the lines of the recent work in Cheng and Escobar-Anel (2023b).

Fig. 7 π versus a

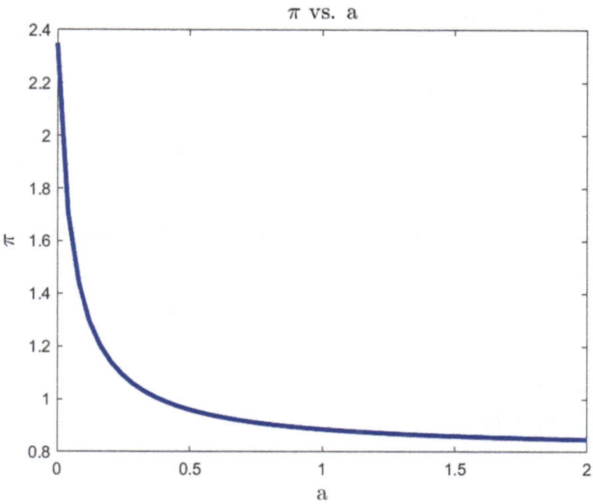

Fig. 8 π versus b

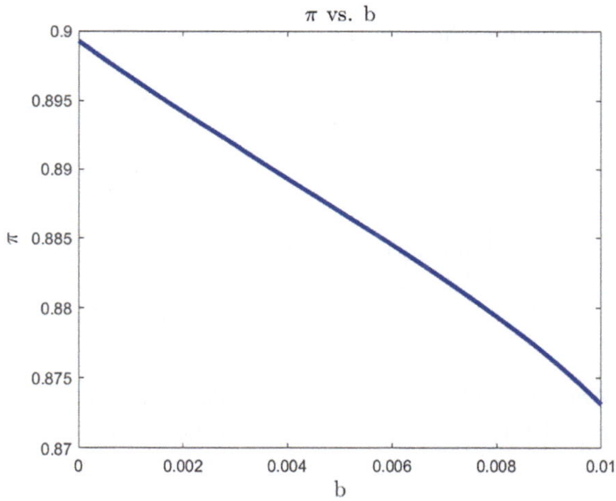

7 Proof of Proposition 3.1 (Incomplete Market, No Robustness, No Consumption)

Proof Solve for $\pi^\star$,

$$xJ_x\left[\rho\bar{\lambda}_1(a\sqrt{v_t}+\frac{b}{\sqrt{v_t}})^2+\sqrt{1-\rho^2}\bar{\lambda}_2(a\sqrt{v_t}+\frac{b}{\sqrt{v_t}})^2\right]+x^2\left(a\sqrt{v}+\frac{b}{\sqrt{v}}\right)^2\pi J_{xx}$$
$$+\sigma x\rho(av+b)J_{xv}=0. \tag{48}$$

It follows that

$$\pi^\star = \frac{xJ_x\left[\rho\bar{\lambda}_1(a\sqrt{v_t}+\frac{b}{\sqrt{v_t}})^2+\sqrt{1-\rho^2}\bar{\lambda}_2(a\sqrt{v_t}+\frac{b}{\sqrt{v_t}})^2\right]+\sigma x\rho(av+b)J_{xv}}{-x^2\left(a\sqrt{v}+\frac{b}{\sqrt{v}}\right)^2 J_{xx}} \tag{49}$$

In the case of no intermediate consumption, we conjecture our value function as follows

$$J(x,v,t)=\frac{x^\gamma}{\gamma}h(t,v), \tag{50}$$

where $h(T,v)=1$ for all v. Thereby, we have following partial derivatives for $J(x,v,t)$:

$$J_t=\frac{x^\gamma}{\gamma}h_t,\ J_v=\frac{x^\gamma}{\gamma}h_v,\ J_x=x^{\gamma-1}h,\ J_{vv}=\frac{x^\gamma}{\gamma}h_{vv},\ J_{xv}=x^{\gamma-1}h_v,\ J_{xx}=(\gamma-1)x^{\gamma-2}h.$$

Substituting these partial derivatives into the optimal allocation $\pi^\star$ from Eq. (49),

$$\begin{aligned}(\pi)^\star &= \frac{x^\gamma h\left[\rho\bar{\lambda}_1(a\sqrt{v_t}+\frac{b}{\sqrt{v_t}})^2+\sqrt{1-\rho^2}\bar{\lambda}_2(a\sqrt{v_t}+\frac{b}{\sqrt{v_t}})^2\right]+\sigma\rho(av+b)x^\gamma h_v}{-(\gamma-1)\left(a\sqrt{v}+\frac{b}{\sqrt{v}}\right)^2 x^\gamma h}\\ &= \frac{\left[\rho\bar{\lambda}_1(a\sqrt{v_t}+\frac{b}{\sqrt{v_t}})^2+\sqrt{1-\rho^2}\bar{\lambda}_2(a\sqrt{v_t}+\frac{b}{\sqrt{v_t}})^2\right]+\sigma\rho(av+b)\frac{h_v}{h}}{-(\gamma-1)\left(a\sqrt{v}+\frac{b}{\sqrt{v}}\right)^2}.\end{aligned} \tag{51}$$

And then substituting the partial derivatives and the optimal allocation above into Eq. (10) to eliminate "*sup*" and further simplify leads to

$$\begin{aligned}&h_t+\left[r\gamma-\frac{\gamma}{2}\frac{\left(\rho\bar{\lambda}_1(a\sqrt{v_t}+\frac{b}{\sqrt{v_t}})+\sqrt{1-\rho^2}\bar{\lambda}_2(a\sqrt{v}+\frac{b}{\sqrt{v}})\right)^2}{(\gamma-1)}\right]h+\left[-\frac{\gamma}{2}\frac{\sigma^2\rho^2 v}{(\gamma-1)}\right]\frac{h_v^2}{h}\\ &+\left[\kappa(\theta-v)-\gamma\frac{\left(\rho\bar{\lambda}_1(a\sqrt{v_t}+\frac{b}{\sqrt{v_t}})+\sqrt{1-\rho^2}\bar{\lambda}_2(a\sqrt{v}+\frac{b}{\sqrt{v}})\right)}{(\gamma-1)}\sigma\rho\sqrt{v}\right]h_v+\frac{1}{2}\sigma^2 v h_{vv}=0.\end{aligned} \tag{52}$$

It can be seen that if we have

$$\rho=0, \tag{53}$$

the PDE above becomes linear:

$$h_t + V(v,t)h + \Gamma(v,t)h_v + \frac{1}{2}\sigma^2 v h_{vv} = 0, \tag{54}$$

where $V(v,t)$ and $\Gamma(v,t)$ are defined in Eq. (13). Further, the coefficients satisfy the conditions of Heath and Schweizer (2000) as per the previous proposition, and h admits the Feynman-Kac representation:

$$\begin{aligned} &h(v,t) \\ &= \mathbb{E}^{Q}\left[\exp\left\{\int_t^T V(v_\tau,\tau)d\tau\right\}h(v,T) \mid v_t\right] \\ &= \exp\left\{\left(r\gamma - \frac{\gamma}{\gamma-1}(\bar{\lambda}_1\rho + \bar{\lambda}_2\sqrt{1-\rho^2})^2 ab\right)(T-t)\right\} \times \underbrace{\mathbb{E}^{Q}\left[\exp\left\{-\mu\int_t^T v_\tau d\tau - \nu\int_t^T \frac{1}{v_\tau}d\tau\right\} \mid v_t\right]}_{\triangleq q(T-t,v;\alpha,\lambda,\mu,\nu)}, \end{aligned} \tag{55}$$

with parameters

$$\begin{aligned} \alpha &= 0, \\ \lambda &= 0, \\ \mu &= \frac{1}{2}\frac{\gamma}{\gamma-1}(\rho\bar{\lambda}_1 + \sqrt{1-\rho^2}\bar{\lambda}_2)^2 a^2, \\ \nu &= \frac{1}{2}\frac{\gamma}{\gamma-1}(\rho\bar{\lambda}_1 + \sqrt{1-\rho^2}\bar{\lambda}_2)^2 b^2. \end{aligned} \tag{56}$$

Note that the conditional expectation is taken under the probability measure $\mathbb{Q}$ such that v_t has drift $\Gamma(v,t)$ in Eq. (54) instead of $\kappa(\theta - v)$. The Feller condition is assumed to be satisfied by the new drift:

$$\begin{aligned} \Gamma(v,t) &= \kappa(\theta - v) - \gamma\frac{\left(\rho\bar{\lambda}_1(a\sqrt{v_t} + \frac{b}{\sqrt{v_t}}) + \sqrt{1-\rho^2}\bar{\lambda}_2(a\sqrt{v} + \frac{b}{\sqrt{v}})\right)}{(\gamma-1)}\sigma\rho\sqrt{v} \\ &= \kappa(\theta - v) - \frac{\gamma}{\gamma-1}\left(\rho\bar{\lambda}_1 + \sqrt{1-\rho^2}\bar{\lambda}_2\right)\sigma\rho(av+b) \\ &= \underbrace{\left(\kappa\theta - \frac{\gamma}{\gamma-1}\left(\rho\bar{\lambda}_1 + \sqrt{1-\rho^2}\bar{\lambda}_2\right)\sigma\rho b\right)}_{\tilde{\kappa}\tilde{\theta}} - \underbrace{\left(\kappa + \frac{\gamma}{\gamma-1}\left(\rho\bar{\lambda}_1 + \sqrt{1-\rho^2}\bar{\lambda}_2\right)\sigma\rho a\right)}_{\tilde{\kappa}} v \\ &\Longrightarrow \kappa\theta - \frac{\gamma}{\gamma-1}\left(\rho\bar{\lambda}_1 + \sqrt{1-\rho^2}\bar{\lambda}_2\right)\sigma\rho b \geq \frac{\sigma^2}{2} \end{aligned} \tag{57}$$

Further, if α, λ, μ, and ν in Eq. (56) satisfy the conditions in Eq. (58):

$$\begin{aligned} \mu &> \frac{-\tilde{\kappa}^2}{2\sigma^2}, \\ \nu &\geq -\frac{(2\tilde{\kappa}\tilde{\theta}-\sigma^2)^2}{8\sigma^2}, \\ \alpha &< \frac{1}{2\sigma^2}\left(2\tilde{\kappa}\tilde{\theta}+\sigma^2\sqrt{(2\tilde{\kappa}\tilde{\theta}-\sigma^2)^2+8\sigma^2\nu}\right), \\ \lambda &\geq -\frac{\sqrt{\tilde{\kappa}^2+2\mu\sigma^2}+\tilde{\kappa}}{\sigma^2}, \end{aligned} \tag{58}$$

then Equation of (17) can be solved explicitly by Grasselli (2017)'s result like Eq. (18) with associated m, D, β, and K like Eq. (19). Note that the last two conditions for $\alpha = \lambda = 0$ are satisfied directly. Thus, the dependence of the function $q(\cdot)$ on α and λ can be omitted.

Moreover, the optimal strategy with $\rho = 0$ is given by

$$\begin{aligned} (\pi)^\star &= \frac{-\sqrt{v}\sigma\rho h_v - \left(\rho\bar{\lambda}_1+\sqrt{1-\rho^2}\bar{\lambda}_2\right)(a\sqrt{v_t}+\frac{b}{\sqrt{v_t}})h}{(a\sqrt{v}+\frac{b}{\sqrt{v}})(\gamma-1)h} \\ &= \frac{-\bar{\lambda}_2(a\sqrt{v_t}+\frac{b}{\sqrt{v_t}})h}{(a\sqrt{v}+\frac{b}{\sqrt{v}})(\gamma-1)h} \\ &= \frac{-\bar{\lambda}_2}{\gamma-1}. \end{aligned} \tag{59}$$

□

8 Proof of Proposition 4.1 (Incomplete Market, Robustness, Consumption)

Proof Following the similar procedure as before, we solve an infimization problem in terms of e^v and e^S first, and then substitute back into the robust HJB equation (28) to solve the maximization problem in terms of the optimal allocation π.

Solving the infimization problem leads to

$$\begin{cases} \frac{e^v}{\Phi_1} = \pi\rho\left(a\sqrt{v}+\frac{b}{\sqrt{v}}\right)xJ_x+\sigma\sqrt{v}J_v \\ \frac{e^S}{\Phi_2} = \pi\sqrt{1-\rho^2}\left(a\sqrt{v}+\frac{b}{\sqrt{v}}\right)xJ_x \end{cases} \Longrightarrow \begin{cases} (e^v)^\star = \Phi_1\left[\pi\rho\left(a\sqrt{v}+\frac{b}{\sqrt{v}}\right)xJ_x+\sigma\sqrt{v}J_v\right] \\ (e^S)^\star = \Phi_2\left[\pi\sqrt{1-\rho^2}\left(a\sqrt{v}+\frac{b}{\sqrt{v}}\right)xJ_x\right]. \end{cases} \tag{60}$$

The HJB equation then follows

$$
\begin{aligned}
&\sup_{\pi,c}\Bigg\{u(c)-\delta J+J_t+\Bigg(x\Bigg[r+\pi\Big[\rho\bar{\lambda}_1(a\sqrt{v_t}+\frac{b}{\sqrt{v_t}})^2+\sqrt{1-\rho^2}\bar{\lambda}_2(a\sqrt{v_t}+\frac{b}{\sqrt{v_t}})^2\\
&\quad-\rho(a\sqrt{v_t}+\frac{b}{\sqrt{v_t}})\Phi_1\Big[\pi\rho\left(a\sqrt{v}+\frac{b}{\sqrt{v}}\right)xJ_x+\sigma\sqrt{v}J_v\Big]\\
&\quad-\sqrt{1-\rho^2}(a\sqrt{v_t}+\frac{b}{\sqrt{v_t}})\Phi_2\Big[\pi\sqrt{1-\rho^2}\left(a\sqrt{v}+\frac{b}{\sqrt{v}}\right)xJ_x\Big]\Big]\Bigg]-c\Bigg)J_x\\
&\quad+\frac{1}{2}x^2\left(a\sqrt{v}+\frac{b}{\sqrt{v}}\right)^2\pi^2J_{xx}\\
&\quad+\Big[\kappa(\theta-v)-\sigma\sqrt{v}\Phi_1\Big[\pi\rho\left(a\sqrt{v}+\frac{b}{\sqrt{v}}\right)xJ_x+\sigma\sqrt{v}J_v\Big]\Big]J_v\\
&\quad+\frac{1}{2}\sigma^2vJ_{vv}+\sigma x\pi\rho(av+b)J_{xv}+\frac{\Phi_1^2\Big[\pi\rho\left(a\sqrt{v}+\frac{b}{\sqrt{v}}\right)xJ_x+\sigma\sqrt{v}J_v\Big]^2}{2\Phi_1}\\
&\quad+\frac{\Phi_2^2\Big[\pi\sqrt{1-\rho^2}\left(a\sqrt{v}+\frac{b}{\sqrt{v}}\right)xJ_x\Big]^2}{2\Phi_2}\Bigg\}=0.
\end{aligned}
\tag{61}
$$

Simplifying leads to

$$
\begin{aligned}
&\sup_{\pi,c}\Bigg\{u(c)-\delta J+J_t+\Bigg(x\Bigg[r+\pi\Big[\rho\bar{\lambda}_1(a\sqrt{v_t}+\frac{b}{\sqrt{v_t}})^2+\sqrt{1-\rho^2}\bar{\lambda}_2(a\sqrt{v_t}+\frac{b}{\sqrt{v_t}})^2\Big]\Bigg]-c\Bigg)J_x\\
&\quad-\Phi_1\Big[\pi^2\rho^2\left(a\sqrt{v}+\frac{b}{\sqrt{v}}\right)^2x^2J_x^2+\pi\sigma\rho(av+b)xJ_xJ_v\Big]\\
&\quad-\Phi_2\Big[\pi^2(1-\rho^2)\left(a\sqrt{v}+\frac{b}{\sqrt{v}}\right)^2x^2J_x^2\Big]+\frac{1}{2}x^2\left(a\sqrt{v}+\frac{b}{\sqrt{v}}\right)^2\pi^2J_{xx}\\
&\quad+\kappa(\theta-v)J_v-\Phi_1\Big[\pi\rho\sigma(av+b)xJ_xJ_v+\sigma^2vJ_v^2\Big]\\
&\quad+\frac{1}{2}\sigma^2vJ_{vv}+\sigma x\pi\rho(av+b)J_{xv}\\
&\quad+\frac{\Phi_1\Big[\pi^2\rho^2\left(a\sqrt{v}+\frac{b}{\sqrt{v}}\right)^2x^2J_x^2+2\pi\rho\sigma(av+b)xJ_xJ_v+\sigma^2vJ_v^2\Big]}{2}\\
&\quad+\frac{\Phi_2\Big[\pi^2(1-\rho^2)\left(a\sqrt{v}+\frac{b}{\sqrt{v}}\right)^2x^2J_x^2\Big]}{2}\Bigg\}=0.
\end{aligned}
\tag{62}
$$

That is,

$$
\begin{aligned}
&\sup_{\pi,c}\Bigg\{u(c)-\delta J+J_t+\Bigg(x\Bigg[r+\pi\Big[\rho\bar{\lambda}_1(a\sqrt{v_t}+\frac{b}{\sqrt{v_t}})^2+\sqrt{1-\rho^2}\bar{\lambda}_2(a\sqrt{v_t}+\frac{b}{\sqrt{v_t}})^2\Big]\Bigg]-c\Bigg)J_x\\
&\quad+\frac{1}{2}x^2\left(a\sqrt{v}+\frac{b}{\sqrt{v}}\right)^2\pi^2J_{xx}+\kappa(\theta-v)J_v+\frac{1}{2}\sigma^2vJ_{vv}+\sigma x\pi\rho(av+b)J_{xv}\\
&\quad-\frac{\Phi_1\Big[\pi^2\rho^2\left(a\sqrt{v}+\frac{b}{\sqrt{v}}\right)^2x^2J_x^2+2\pi\rho\sigma(av+b)xJ_xJ_v+\sigma^2vJ_v^2\Big]}{2}\\
&\quad-\frac{\Phi_2\Big[\pi^2(1-\rho^2)\left(a\sqrt{v}+\frac{b}{\sqrt{v}}\right)^2x^2J_x^2\Big]}{2}\Bigg\}=0.
\end{aligned}
\tag{63}
$$

Solve for $c^\star$,

$$\begin{aligned} 0 &= u'(c) - J_x \\ &= \epsilon_1 c^{\gamma-1} - J_x \end{aligned} \tag{64}$$

It follows that

$$c^\star = (J_x)^{\frac{1}{\gamma-1}}(\epsilon_1)^{-\frac{1}{\gamma-1}}. \tag{65}$$

Solve for $\pi^\star$,

$$\begin{aligned} & xJ_x\Big[\rho\bar{\lambda}_1(a\sqrt{v_t}+\frac{b}{\sqrt{v_t}})^2+\sqrt{1-\rho^2}\bar{\lambda}_2(a\sqrt{v_t}+\frac{b}{\sqrt{v_t}})^2\Big]+x^2\left(a\sqrt{v}+\frac{b}{\sqrt{v}}\right)^2\pi J_{xx} \\ & +\sigma x\rho(av+b)J_{xv}-\Phi_1\Big[\pi\rho^2\left(a\sqrt{v}+\frac{b}{\sqrt{v}}\right)^2x^2J_x^2+\rho\sigma(av+b)xJ_xJ_v\Big] \\ & -\Phi_2\Big[\pi(1-\rho^2)\left(a\sqrt{v}+\frac{b}{\sqrt{v}}\right)^2x^2J_x^2\Big] \\ & =0. \end{aligned} \tag{66}$$

It follows that

$$\pi^\star = \frac{xJ_x\Big[\rho\bar{\lambda}_1(a\sqrt{v_t}+\frac{b}{\sqrt{v_t}})^2+\sqrt{1-\rho^2}\bar{\lambda}_2(a\sqrt{v_t}+\frac{b}{\sqrt{v_t}})^2\Big]+\sigma x\rho(av+b)J_{xv}-\Phi_1\rho\sigma(av+b)xJ_xJ_v}{\Phi_1\rho^2\left(a\sqrt{v}+\frac{b}{\sqrt{v}}\right)^2x^2J_x^2+\Phi_2(1-\rho^2)\left(a\sqrt{v}+\frac{b}{\sqrt{v}}\right)^2x^2J_x^2-x^2\left(a\sqrt{v}+\frac{b}{\sqrt{v}}\right)^2J_{xx}}. \tag{67}$$

In the case of considering both intermediate consumption and terminal wealth, we conjecture our value function has the following form

$$J(x,v,t) = \frac{x^\gamma}{\gamma}\Big(h(t,v)\Big)^{1-\gamma}, \tag{68}$$

where $h(T,v)=1$ for all v. Thereby, we have following partial derivatives for $J(x,v,t)$:

$$\begin{aligned} J_t &= \frac{1-\gamma}{\gamma}x^\gamma h^{-\gamma}h_t, \\ J_v &= \frac{x^\gamma}{\gamma}(1-\gamma)h^{-\gamma}h_v, \\ J_x &= x^{\gamma-1}h^{1-\gamma}, \\ J_{vv} &= -(1-\gamma)x^\gamma h^{-\gamma-1}h_v^2+\frac{x^\gamma}{\gamma}(1-\gamma)h^{-\gamma}h_{vv}, \\ J_{xv} &= (1-\gamma)x^{\gamma-1}h^{-\gamma}h_v, \\ J_{xx} &= (\gamma-1)x^{\gamma-2}h^{1-\gamma}. \end{aligned}$$

Substituting these partial derivatives into the candidates for the optimal consumption in Eq. (65) and optimal allocation in Eq. (67) with $\Phi_i = \frac{\phi_i}{\gamma J} = \frac{\phi_i}{x^\gamma h^{1-\gamma}}$ as in Maenhout (2004) into the above equation, we have

$$
\begin{cases}
c^{\star} &= x h^{-1}(\epsilon_1)^{-\frac{1}{\gamma-1}} \\
\pi^{\star} &= \dfrac{x^{\gamma}h^{1-\gamma}\left[\rho\bar{\lambda}_1(a\sqrt{v_t}+\frac{b}{\sqrt{v_t}})^2+\sqrt{1-\rho^2}\bar{\lambda}_2(a\sqrt{v_t}+\frac{b}{\sqrt{v_t}})^2\right]+\sigma\rho(av+b)(1-\gamma)x^{\gamma}h^{-\gamma}h_v-\frac{\phi_1}{x^{\gamma}h^{1-\gamma}}\rho\sigma(av+b)\frac{1-\gamma}{\gamma}x^{2\gamma}h^{1-2\gamma}h_v}{\frac{\phi_1}{x^{\gamma}h^{1-\gamma}}\rho^2\left(a\sqrt{v}+\frac{b}{\sqrt{v}}\right)^2x^{2\gamma}h^{2-2\gamma}+\frac{\phi_2}{x^{\gamma}h^{1-\gamma}}(1-\rho^2)\left(a\sqrt{v}+\frac{b}{\sqrt{v}}\right)^2x^{2\gamma}h^{2-2\gamma}-\left(a\sqrt{v}+\frac{b}{\sqrt{v}}\right)^2(\gamma-1)x^{\gamma}h^{1-\gamma}} \\
&= \dfrac{x^{\gamma}h^{1-\gamma}\left[\rho\bar{\lambda}_1(a\sqrt{v_t}+\frac{b}{\sqrt{v_t}})^2+\sqrt{1-\rho^2}\bar{\lambda}_2(a\sqrt{v_t}+\frac{b}{\sqrt{v_t}})^2\right]+\sigma\rho(av+b)(1-\gamma)x^{\gamma}h^{-\gamma}h_v-\phi_1\rho\sigma(av+b)\frac{1-\gamma}{\gamma}x^{\gamma}h^{-\gamma}h_v}{\phi_1\rho^2\left(a\sqrt{v}+\frac{b}{\sqrt{v}}\right)^2x^{\gamma}h^{1-\gamma}+\phi_2(1-\rho^2)\left(a\sqrt{v}+\frac{b}{\sqrt{v}}\right)^2x^{\gamma}h^{1-\gamma}-\left(a\sqrt{v}+\frac{b}{\sqrt{v}}\right)^2(\gamma-1)x^{\gamma}h^{1-\gamma}} \\
&= \dfrac{x^{\gamma}h^{1-\gamma}\left(\rho\bar{\lambda}_1+\sqrt{1-\rho^2}\bar{\lambda}_2\right)(a\sqrt{v_t}+\frac{b}{\sqrt{v_t}})^2+(1-\frac{\phi_1}{\gamma})\sigma\rho(av+b)(1-\gamma)x^{\gamma}h^{-\gamma}h_v}{\left(\phi_1\rho^2+\phi_2(1-\rho^2)-(\gamma-1)\right)\left(a\sqrt{v}+\frac{b}{\sqrt{v}}\right)^2x^{\gamma}h^{1-\gamma}} \\
&= \dfrac{h^{1-\gamma}\left(\rho\bar{\lambda}_1+\sqrt{1-\rho^2}\bar{\lambda}_2\right)(a\sqrt{v_t}+\frac{b}{\sqrt{v_t}})^2+(1-\frac{\phi_1}{\gamma})\sigma\rho(av+b)(1-\gamma)h^{-\gamma}h_v}{\left(\phi_1\rho^2+\phi_2(1-\rho^2)-(\gamma-1)\right)\left(a\sqrt{v}+\frac{b}{\sqrt{v}}\right)^2h^{1-\gamma}}
\end{cases}
\quad (69)
$$

Substituting the partial derivatives and the optimal allocation in Equation above into Eq. (63) to eliminate "sup", and simplifying the equation lead to

$$
\begin{aligned}
&(1-\gamma)(\epsilon_1)^{-\frac{1}{\gamma-1}}-\delta h+(1-\gamma)h_t+r\gamma h \\
&+\frac{(\rho\bar{\lambda}_1+\sqrt{1-\rho^2}\bar{\lambda}_2)^2(a\sqrt{v}+\frac{b}{\sqrt{v}})^2}{\left(\phi_1\rho^2+\phi_2(1-\rho^2)-(\gamma-1)\right)}\gamma h+\frac{(1-\frac{\phi_1}{\gamma})(1-\gamma)\sigma\rho(av+b)(\rho\bar{\lambda}_1+\sqrt{1-\rho^2}\bar{\lambda}_2)}{\left(\phi_1\rho^2+\phi_2(1-\rho^2)-(\gamma-1)\right)}\gamma h_v \\
&+\frac{1}{2}(\gamma-1)\Bigg[\frac{(\rho\bar{\lambda}_1+\sqrt{1-\rho^2}\bar{\lambda}_2)^2(a\sqrt{v}+\frac{b}{\sqrt{v}})^2}{\left(\phi_1\rho^2+\phi_2(1-\rho^2)-(\gamma-1)\right)^2}\gamma h+2\frac{(\rho\bar{\lambda}_1+\sqrt{1-\rho^2}\bar{\lambda}_2)(av+b)(1-\frac{\phi_1}{\gamma})(1-\gamma)}{\left(\phi_1\rho^2+\phi_2(1-\rho^2)-(\gamma-1)\right)^2}\gamma h_v \\
&+\frac{(1-\frac{\phi_1}{\gamma})^2(1-\gamma)^2\sigma^2\rho^2 v}{\left(\phi_1\rho^2+\phi_2(1-\rho^2)-(\gamma-1)\right)^2}\gamma\frac{h_v^2}{h}\Bigg] \\
&+\kappa(\theta-v)(1-\gamma)h_v+\frac{1}{2}\sigma^2 v(1-\gamma)\gamma\frac{h_v^2}{h}+\frac{1}{2}\sigma^2 v(1-\gamma)h_{vv} \\
&+\frac{\sigma\rho(1-\gamma)(\rho\bar{\lambda}_1+\sqrt{1-\rho^2}\bar{\lambda}_2)(av+b)}{\left(\phi_1\rho^2+\phi_2(1-\rho^2)-(\gamma-1)\right)}\gamma h_v+\frac{\sigma^2\rho^2(1-\frac{\phi_1}{\gamma})(1-\gamma)^2 v}{\left(\phi_1\rho^2+\phi_2(1-\rho^2)-(\gamma-1)\right)}\gamma\frac{h_v^2}{h} \\
&-\frac{\phi_1}{2}\Bigg\{\frac{\rho^2(\rho\bar{\lambda}_1+\sqrt{1-\rho^2}\bar{\lambda}_2)^2(a\sqrt{v}+\frac{b}{\sqrt{v}})^2}{\left(\phi_1\rho^2+\phi_2(1-\rho^2)-(\gamma-1)\right)^2}\gamma h+2\frac{\rho^2(\rho\bar{\lambda}_1+\sqrt{1-\rho^2}\bar{\lambda}_2)(av+b)(1-\frac{\phi_1}{\gamma})(1-\gamma)\sigma\rho}{\left(\phi_1\rho^2+\phi_2(1-\rho^2)-(\gamma-1)\right)^2}\gamma h_v \\
&+\frac{\rho^4(1-\frac{\phi_1}{\gamma})^2(1-\gamma)^2\sigma^2 v}{\left(\phi_1\rho^2+\phi_2(1-\rho^2)-(\gamma-1)\right)^2}\gamma\frac{h_v^2}{h}+2\sigma\rho\frac{(\rho\bar{\lambda}_1+\sqrt{1-\rho^2}\bar{\lambda}_2)(av+b)}{\left(\phi_1\rho^2+\phi_2(1-\rho^2)-(\gamma-1)\right)}(1-\gamma)h_v \\
&+2\sigma^2\rho^2\frac{(1-\frac{\phi_1}{\gamma})(1-\gamma)v}{\left(\phi_1\rho^2+\phi_2(1-\rho^2)-(\gamma-1)\right)}(1-\gamma)\frac{h_v^2}{h}+\sigma^2 v(1-\gamma)^2\frac{h_v^2}{h}\Bigg\} \\
&-\frac{\phi_2}{2}\Bigg\{\frac{(1-\rho^2)(\rho\bar{\lambda}_1+\sqrt{1-\rho^2}\bar{\lambda}_2)^2(a\sqrt{v}+\frac{b}{\sqrt{v}})^2}{\left(\phi_1\rho^2+\phi_2(1-\rho^2)-(\gamma-1)\right)^2}\gamma h+2\frac{(1-\rho^2)(\rho\bar{\lambda}_1+\sqrt{1-\rho^2}\bar{\lambda}_2)(av+b)(1-\frac{\phi_1}{\gamma})(1-\gamma)\sigma\rho}{\left(\phi_1\rho^2+\phi_2(1-\rho^2)-(\gamma-1)\right)^2}\gamma h_v \\
&+\frac{(1-\rho^2)(1-\frac{\phi_1}{\gamma})^2(1-\gamma)^2\sigma^2\rho^2 v}{\left(\phi_1\rho^2+\phi_2(1-\rho^2)-(\gamma-1)\right)^2}\gamma\frac{h_v^2}{h}\Bigg\}=0.
\end{aligned}
\quad (70)
$$

Regrouping terms leads to

$$
\begin{aligned}
&(1-\gamma)(\epsilon_1)^{-\frac{1}{\gamma-1}} + (1-\gamma)h_t \\
&+\Bigg[r - \frac{\delta}{\gamma} + \frac{(\rho\bar{\lambda}_1 + \sqrt{1-\rho^2}\bar{\lambda}_2)^2(a\sqrt{v} + \frac{b}{\sqrt{v}})^2}{\Big(\phi_1\rho^2 + \phi_2(1-\rho^2) - (\gamma-1)\Big)} + \frac{1}{2}(\gamma-1)\frac{(\rho\bar{\lambda}_1 + \sqrt{1-\rho^2}\bar{\lambda}_2)^2(a\sqrt{v} + \frac{b}{\sqrt{v}})^2}{\Big(\phi_1\rho^2 + \phi_2(1-\rho^2) - (\gamma-1)\Big)^2} \\
&- \frac{1}{2}(\phi_1\rho^2 + \phi_2(1-\rho^2))\frac{(\rho\bar{\lambda}_1 + \sqrt{1-\rho^2}\bar{\lambda}_2)^2(a\sqrt{v} + \frac{b}{\sqrt{v}})^2}{\Big(\phi_1\rho^2 + \phi_2(1-\rho^2) - (\gamma-1)\Big)^2}\Bigg]\gamma h \\
&+\Bigg[\frac{(1-\frac{\phi_1}{\gamma})(1-\gamma)\sigma\rho(av+b)(\rho\bar{\lambda}_1 + \sqrt{1-\rho^2}\bar{\lambda}_2)}{\Big(\phi_1\rho^2 + \phi_2(1-\rho^2) - (\gamma-1)\Big)}\gamma + \frac{(\rho\bar{\lambda}_1 + \sqrt{1-\rho^2}\bar{\lambda}_2)(av+b)(1-\frac{\phi_1}{\gamma})(1-\gamma)}{\Big(\phi_1\rho^2 + \phi_2(1-\rho^2) - (\gamma-1)\Big)^2}\gamma(\gamma-1) \\
&+ \kappa(\theta - v)(1-\gamma) + \frac{\sigma\rho(1-\gamma)(\rho\bar{\lambda}_1 + \sqrt{1-\rho^2}\bar{\lambda}_2)(av+b)}{\Big(\phi_1\rho^2 + \phi_2(1-\rho^2) - (\gamma-1)\Big)}\gamma \\
&- (\phi_1\rho^2 + \phi_2(1-\rho^2))\frac{(\rho\bar{\lambda}_1 + \sqrt{1-\rho^2}\bar{\lambda}_2)(av+b)(1-\frac{\phi_1}{\gamma})(1-\gamma)\sigma\rho}{\Big(\phi_1\rho^2 + \phi_2(1-\rho^2) - (\gamma-1)\Big)^2}\gamma \\
&- \phi_1\sigma\rho\frac{(\rho\bar{\lambda}_1 + \sqrt{1-\rho^2}\bar{\lambda}_2)(av+b)}{\Big(\phi_1\rho^2 + \phi_2(1-\rho^2) - (\gamma-1)\Big)}(1-\gamma)\Bigg]h_v \\
&+\Bigg[\frac{1}{2}\frac{(1-\frac{\phi_1}{\gamma})^2(1-\gamma)^2\sigma^2\rho^2 v}{\Big(\phi_1\rho^2 + \phi_2(1-\rho^2) - (\gamma-1)\Big)^2}\gamma(\gamma-1) + \frac{1}{2}\sigma^2 v(1-\gamma)\gamma \\
&+ \frac{\sigma^2\rho^2(1-\frac{\phi_1}{\gamma})(1-\gamma)^2 v}{\Big(\phi_1\rho^2 + \phi_2(1-\rho^2) - (\gamma-1)\Big)}\gamma - \frac{1}{2}(\phi_1\rho^2 + \phi_2(1-\rho^2))\frac{(1-\frac{\phi_1}{\gamma})^2(1-\gamma)^2\sigma^2\rho^2 v}{\Big(\phi_1\rho^2 + \phi_2(1-\rho^2) - (\gamma-1)\Big)^2}\gamma \\
&- \phi_1\sigma^2\rho^2\frac{(1-\frac{\phi_1}{\gamma})(1-\gamma)v}{\Big(\phi_1\rho^2 + \phi_2(1-\rho^2) - (\gamma-1)\Big)}(1-\gamma) - \frac{\phi_1}{2}\sigma^2 v(1-\gamma)^2\Bigg]\frac{h_v^2}{h} \\
&+\Big[\frac{1}{2}\sigma^2 v(1-\gamma)\Big]h_{vv} = 0.
\end{aligned}
\tag{71}
$$

That is,

$$
\begin{aligned}
&(1-\gamma)(\epsilon_1)^{-\frac{1}{\gamma-1}} + (1-\gamma)h_t \\
&+\left[r-\frac{\delta}{\gamma}+\frac{1}{2}\frac{(\rho\bar{\lambda}_1+\sqrt{1-\rho^2}\bar{\lambda}_2)^2(a\sqrt{v}+\frac{b}{\sqrt{v}})^2}{\left(\phi_1\rho^2+\phi_2(1-\rho^2)-(\gamma-1)\right)}\right]\gamma h \\
&+\left[\kappa(\theta-v)(1-\gamma)+\frac{\sigma\rho(1-\gamma)(\rho\bar{\lambda}_1+\sqrt{1-\rho^2}\bar{\lambda}_2)(av+b)}{\left(\phi_1\rho^2+\phi_2(1-\rho^2)-(\gamma-1)\right)}(\gamma+\phi_1)\right]h_v \\
&+\left[\frac{1}{2}\sigma^2 v(1-\gamma)(\gamma-\phi_1(1-\gamma))+\frac{1}{2}(\gamma-\phi_1)\frac{\sigma^2\rho^2(1-\frac{\phi_1}{\gamma})(1-\gamma)^2 v}{\left(\phi_1\rho^2+\phi_2(1-\rho^2)-(\gamma-1)\right)}\right]\frac{h_v^2}{h} \\
&+\left[\frac{1}{2}\sigma^2 v(1-\gamma)\right]h_{vv}=0.
\end{aligned}
\tag{72}
$$

We can rewrite it as:

$$
\begin{aligned}
&(\epsilon_1)^{-\frac{1}{\gamma-1}}+h_t+\underbrace{\left[r-\frac{\delta}{\gamma}+\frac{1}{2}\frac{(\rho\bar{\lambda}_1+\sqrt{1-\rho^2}\bar{\lambda}_2)^2(a\sqrt{v}+\frac{b}{\sqrt{v}})^2}{\left(\phi_1\rho^2+\phi_2(1-\rho^2)-(\gamma-1)\right)}\right]\frac{\gamma}{1-\gamma}}_{\triangleq V(v,t)}h \\
&+\underbrace{\left[\kappa(\theta-v)+\frac{\sigma\rho(\rho\bar{\lambda}_1+\sqrt{1-\rho^2}\bar{\lambda}_2)(av+b)}{\left(\phi_1\rho^2+\phi_2(1-\rho^2)-(\gamma-1)\right)}(\gamma+\phi_1)\right]}_{\triangleq\Gamma(v,t)}h_v+\frac{1}{2}\sigma^2 v h_{vv}=0.
\end{aligned}
\tag{73}
$$

where we have eliminated the non-linear term by enforcing:

$$
\gamma-\phi_1(1-\gamma)=\frac{-\rho^2(\gamma-\phi_1)^2(1-\gamma)}{\gamma\left(\phi_1\rho^2+\phi_2(1-\rho^2)-(\gamma-1)\right)}. \tag{74}
$$

Then we have a linear PDE

$$
(\epsilon_1)^{-\frac{1}{\gamma-1}}+h_t+V(v,t)h+\Gamma(v,t)h_v+\frac{1}{2}\sigma^2 v h_{vv}=0. \tag{75}
$$

Further, the coefficients in (31) satisfy the condition of Heath and Schweizer (2000) as per previous proposition, then h admits the Feynman-Kac representation:

$$
\begin{aligned}
&h(v,t) \\
&=\mathbb{E}^{Q}\left[\int_t^T \exp\left\{\int_t^{\tilde{\tau}} V(v_\tau,\tau)d\tau\right\}\left((\epsilon_1)^{-\frac{1}{\gamma-1}}\right)d\tilde{\tau}+\exp\left\{\int_t^T V(v_\tau,\tau)d\tau\right\}h(v,T)\mid v_t\right].
\end{aligned}
\tag{76}
$$

Moreover, given that $V(v,t)$: $\mathbb{R}^+ \times [0,T] \longrightarrow \mathbb{R}$ is a measurable function, such that $\exp\left\{\int_t^T V(v_\tau,\tau)d\tau\right\} \geq 0$ we can then apply Fubini's theorem, in particular the version of Tonelli's Theorem, to the first term:

$$\begin{aligned} h(v,t) &= \int_t^T \mathbb{E}^Q\Big[\Big((\epsilon_1)^{-\frac{1}{\gamma-1}}\Big)\exp\Big\{\int_t^{\tilde{\tau}} V(v_\tau,\tau)d\tau\Big\} \mid v_t\Big]d\tilde{\tau} + \mathbb{E}^Q\Big[\exp\Big\{\int_t^T V(v_\tau,\tau)d\tau\Big\} \mid v_t\Big] \\ &= (\epsilon_1)^{-\frac{1}{\gamma-1}} \int_t^T \underbrace{\mathbb{E}^Q\Big[\exp\Big\{\int_t^{\tilde{\tau}} V(v_\tau,\tau)d\tau\Big\} \mid v_t\Big]}_{\triangleq g(v,\tilde{\tau})} d\tilde{\tau} + \underbrace{\mathbb{E}^Q\Big[\exp\Big\{\int_t^T V(v_\tau,\tau)d\tau\Big\} \mid v_t\Big]}_{\triangleq g(v,\tau)}. \end{aligned} \tag{77}$$

Here, $\tau(t) = T - t$ and $g(v,\tau)$ can be rewritten as

$$\begin{aligned} g(v,\tau) &= \mathbb{E}^Q\Big[\exp\Big\{\int_t^T V(v_\tau,\tau)d\tau\Big\} \mid v_t\Big] \\ &= \exp\Big\{\frac{\gamma}{1-\gamma}\Big(r - \frac{\delta}{\gamma} - \frac{(\bar{\lambda}_1\rho + \bar{\lambda}_2\sqrt{1-\rho^2})^2 ab}{(\gamma-1) - \phi_1\rho^2 - \phi_2(1-\rho^2)}\Big)(T-t)\Big\} \times \underbrace{\mathbb{E}^Q\Big[\exp\Big\{-\mu\int_t^T v_\tau d\tau - \nu\int_t^T \frac{1}{v_\tau}d\tau\Big\} \mid v_t\Big]}_{\triangleq q(T-t,v;\alpha,\lambda,\mu,\nu)}, \end{aligned} \tag{78}$$

with parameters

$$\begin{aligned} \alpha &= 0, \\ \lambda &= 0, \\ \mu &= \frac{1}{2}\frac{\gamma}{1-\gamma}\Bigg(\frac{(\bar{\lambda}_1\rho + \bar{\lambda}_2\sqrt{1-\rho^2})^2}{(\gamma-1) - \phi_1\rho^2 - \phi_2(1-\rho^2)}\Bigg)a^2, \\ \nu &= \frac{1}{2}\frac{\gamma}{1-\gamma}\Bigg(\frac{(\bar{\lambda}_1\rho + \bar{\lambda}_2\sqrt{1-\rho^2})^2}{(\gamma-1) - \phi_1\rho^2 - \phi_2(1-\rho^2)}\Bigg)b^2. \end{aligned} \tag{79}$$

Note that the conditional expectation in $g(v,t)$ is taken under the probability measure $\mathbb{Q}$ such that v_t has drift $\Gamma(v,t)$ instead of $\kappa(\theta - v)$. The Feller condition is assumed to be satisfied by the new drift:

$$\begin{aligned} \Gamma(v,t) &= \kappa(\theta - v) + \frac{\sigma\rho(\rho\bar{\lambda}_1 + \sqrt{1-\rho^2}\bar{\lambda}_2)(av+b)}{\Big(\phi_1\rho^2 + \phi_2(1-\rho^2) - (\gamma-1)\Big)}(\gamma+\phi_1) \\ &= \Bigg(\frac{\sigma\rho(\rho\bar{\lambda}_1 + \sqrt{1-\rho^2}\bar{\lambda}_2)b}{\Big(\phi_1\rho^2 + \phi_2(1-\rho^2) - (\gamma-1)\Big)}(\gamma+\phi_1) + \kappa\theta\Bigg) - \Bigg(-\frac{\sigma\rho(\rho\bar{\lambda}_1 + \sqrt{1-\rho^2}\bar{\lambda}_2)a}{\Big(\phi_1\rho^2 + \phi_2(1-\rho^2) - (\gamma-1)\Big)}(\gamma+\phi_1) + \kappa\Bigg)v \\ \Longrightarrow & \frac{\sigma\rho(\rho\bar{\lambda}_1 + \sqrt{1-\rho^2}\bar{\lambda}_2)b}{\Big(\phi_1\rho^2 + \phi_2(1-\rho^2) - (\gamma-1)\Big)}(\gamma+\phi_1) + \kappa\theta \geq \frac{\sigma^2}{2}. \end{aligned} \tag{80}$$

Further, if α, λ, μ, and ν in Eq. (79) satisfy the conditions (58), h can be solved explicitly by Grasselli (2017)'s result like Eq. (18) with associated m, D, β, and K like Equation (19). Note that the last two conditions for $\alpha = \lambda = 0$ are satisfied directly. Thus, the dependence of the function $q(\cdot)$ on α and λ can be omitted. The optimal portfolio and consumption-wealth ratio are given by

$$\begin{cases} \left(\frac{c}{x}\right)^{\star} & = h^{-1}(\epsilon_1)^{-\frac{1}{\gamma-1}} \\ (\pi)^{\star} & = \dfrac{h\left(\rho\bar{\lambda}_1+\sqrt{1-\rho^2}\bar{\lambda}_2\right)(a\sqrt{v_t}+\frac{b}{\sqrt{v_t}})^2+(1-\frac{\phi_1}{\gamma})\sigma\rho(av+b)(1-\gamma)h_v}{\left(\phi_1\rho^2+\phi_2(1-\rho^2)-(\gamma-1)\right)\left(a\sqrt{v}+\frac{b}{\sqrt{v}}\right)^2 h} \\ & = \dfrac{\rho\bar{\lambda}_1+\sqrt{1-\rho^2}\bar{\lambda}_2}{\phi_1\rho^2+\phi_2(1-\rho^2)-(\gamma-1)} + \dfrac{(1-\frac{\phi_1}{\gamma})\sigma\rho\sqrt{v}(1-\gamma)}{(\phi_1\rho^2+\phi_2(1-\rho^2)-(\gamma-1))\left(a\sqrt{v}+\frac{b}{\sqrt{v}}\right)}\dfrac{h_v}{h} \end{cases} . \tag{81}$$

The worst case measure is determined by

$$\begin{cases} (e^{v})^{\star} & = \phi_1\left[(\pi)^{\star}\rho\left(a\sqrt{v}+\frac{b}{\sqrt{v}}\right)+\sigma\sqrt{v}\frac{1-\gamma}{\gamma}\frac{h_v}{h}\right] \\ (e^{S})^{\star} & = \phi_2\left[(\pi)^{\star}\sqrt{1-\rho^2}\left(a\sqrt{v}+\frac{b}{\sqrt{v}}\right)\right]. \end{cases} \tag{82}$$

□

9 Extra Plots for Section 5.1

The plots of the optimal consumption-wealth ratio c/x versus the premiums of market price of risk $\bar{\lambda}_1$, $\bar{\lambda}_2$, and the relative importance ratio ϵ_1 are presented in Figs. 9, 10, and 11 respectively.

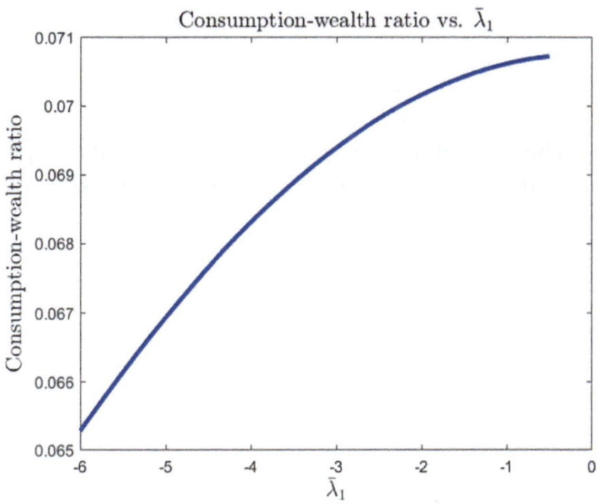

Fig. 9 $\frac{c}{x}$ versus $\bar{\lambda}_1$

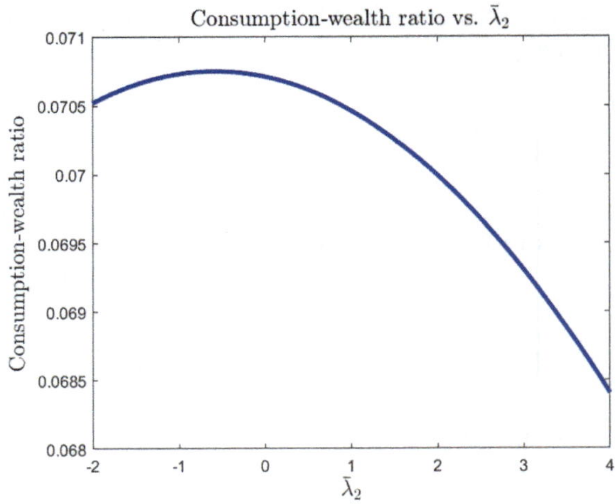

Fig. 10 $\frac{c}{x}$ vs. $\bar{\lambda}_2$

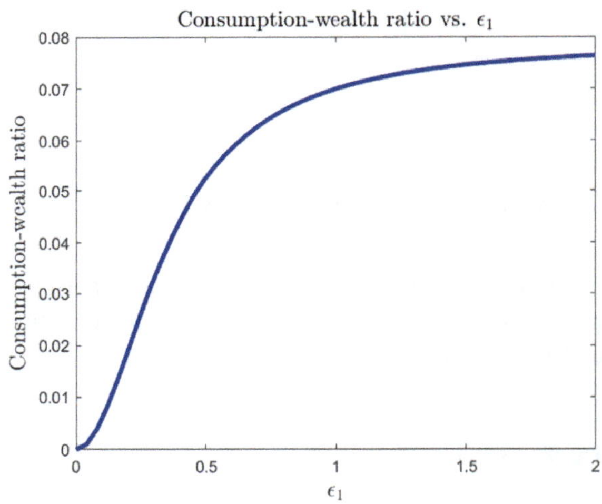

Fig. 11 $\frac{c}{x}$ vs. ϵ_1

The plots of the optimal allocation π versus the premiums of market price of risk $\bar{\lambda}_1$, and $\bar{\lambda}_2$ are presented in Figs. 12, and 13 respectively.

Fig. 12 π vs. $\bar{\lambda}_1$

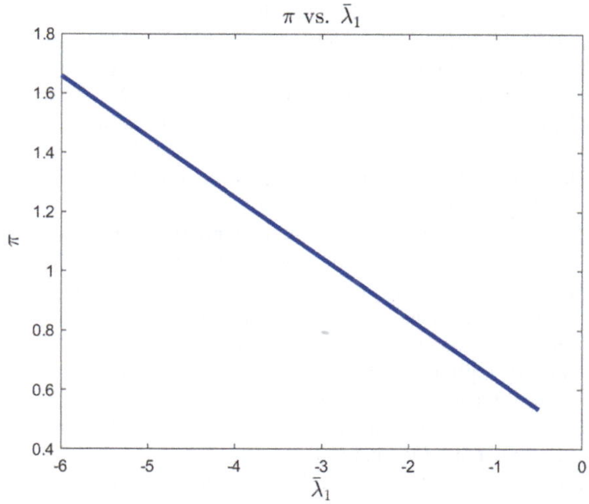

Fig. 13 π vs. $\bar{\lambda}_2$

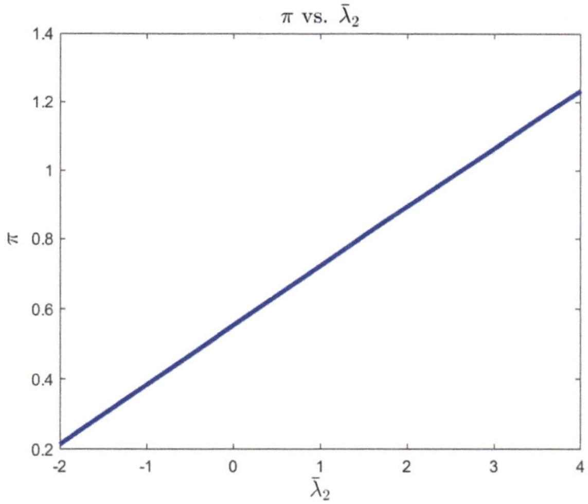

References

Chacko, G., Viceira, L.M.: Dynamic consumption and portfolio choice with stochastic volatility in incomplete markets. Rev. Financ. Stud. **18**(4):1369–1402 (2005)

Cheng, Y., Escobar-Anel, M.: Optimal investment strategy in the family of 4/2 stochastic volatility models. Quant. Finance **21**(10), 1723–1751 (2021)

Cheng, Y., Escobar-Anel, M.: A class of portfolio optimization solvable problems. Financ. Res. Lett. **52**, 103373 (2023)

Cheng, Y., Escobar-Anel, M.: A multivariate 4/2 stochastic covariance model: properties and applications to portfolio decisions. Quant. Finance **23**(3), 497–519 (2023)

Cheng, Y., Escobar-Anel, M.: Optimal consumption and robust portfolio choice for the 3/2 and 4/2 stochastic volatility models. Mathematics **11**(18), 4020 (2023)

Cheng, Y., Escobar-Anel, M.: Robust portfolio choice under the 4/2 stochastic volatility model. IMA J. Manag. Math. **34**(1), 221–256 (2023)
Escobar, M., Ferrando, S., Rubtsov, A.: Robust portfolio choice with derivative trading under stochastic volatility. J. Banking Finance **61**, 142–157 (2015)
Grasselli, M.: The 4/2 stochastic volatility model: a unified approach for the Heston and the 3/2 model. Math. Financ. **27**(4), 1013–1034 (2017)
Heath, D., Schweizer, M.: Martingales versus PDEs in finance: an equivalence result with examples. J. Appl. Probab. 947–957 (2000)
Heston, S.L.: A closed-form solution for options with stochastic volatility with applications to bond and currency options. Rev. Financ. Stud. **6**(2):327–343 (1993)
Heston, S.L.: A simple new formula for options with stochastic volatility (1997)
Jiangyan, P., Zhang, Q.: Robust consumption portfolio optimization with stochastic differential utility. Automatica **133**, 109835 (2021)
Kraft, H.: Optimal portfolios and Heston's stochastic volatility model: an explicit solution for power utility. Quant. Finance **5**(3), 303–313 (2005)
Liu, H.: Robust consumption and portfolio choice for time varying investment opportunities. Ann. Finance **6**(4), 435–454 (2010)
Maenhout, P.J.: Robust portfolio rules and asset pricing. Rev. Financ. Stud. **17**(4):951–983 (2004)
Maenhout, P.J.: Robust portfolio rules and detection-error probabilities for a mean-reverting risk premium. J. Econ. Theory **128**(1):136–163 (2006)
Merton, R.C.: On estimating the expected return on the market: an exploratory investigation. J. Financ. Econ. **8**(4):323–361 (1980)

Characterisation of Asymptotic Behaviour of Perturbed Deterministic and Stochastic Pantograph Equations

John A. D. Appleby and Emmet Lawless

Abstract This paper considers the asymptotic behaviour of deterministically and stochastically forced linear pantograph equations. The asymptotic behaviour is studied in the case when all solutions of the pantograph equation without forcing tend to a trivial equilibrium. In all cases, we give necessary and sufficient conditions on the forcing terms which enable all solutions to converge to the equilibrium of the unforced equation, and which enable solutions to remain bounded. In the deterministic case, we give sharp conditions on forcing terms which enable the solutions of the forced equations to inherit the power law behaviour of the unforced equation, as well as slower rates of decay or growth not present in the unforced equation. Extensions to equations with general unbounded delay, and to finite–dimensional equations are also presented.

Keywords Delay differential equation · Stochastic delay differential equation · Pantograph equation · Asymptotic stability · Polynomial asymptotic stability · Regular variation

1 Motivation and Scope of the Work

Let $a \neq 0, b$ be real numbers, and let $q \in (0, 1)$. The (scalar valued solution of the) delay differential equation

$$u'(t) = bu(t) + au(qt), \quad t \geq 0; \quad u(0) \neq 0, \tag{1}$$

J. A. D. Appleby (✉)
School of Mathematical Sciences, Dublin City University, Dublin, Ireland
e-mail: john.appleby@dcu.ie

E. Lawless
Department of Mathematics, University of Michigan, Ann Arbor, MI, USA
e-mail: elawless@umich.edu

E. Braverman and A. Swishchuk (eds.), *Delay and Stochastic Differential Equations*, ICIAM2023 Springer Series 12, https://doi.org/10.1007/978-981-95-7310-3_7

has attracted a lot of attention. The is a so–called pantograph equation, first studied in depth by Kato and McLeod (1971) and Fox et al. (1971); Ockendon and Tayler (1971), with finite dimensional analysis in Carr and Dyson (1974, 1975). Apart from interest in applications, it attracts mathematical interest as it is an example of a delay differential equation with unbounded delay, and, despite the constant multipliers a and b, the equation is also non–autonomous, by virtue of the "proportional" delay qt. One interesting feature of linear differential equations with unbounded delay is that solutions that tend to an equilibrium need not do so at an exponentially fast rate (see e.g., Appleby and Buckwar 2010; Krisztin 1988; Makay and Terjéki 1998; Diblík 1998; Diblík and Růžičková 2007). Pantograph equations are no exception, and Kato and McLeod (1971) showed that solutions which tend to zero do so at a power law rate of decay. Further treatments of deterministic equations with unbounded delay include Čermák (2002), Haddock and Krisztin (1986), Iserles (1993), Liu (n.d.), and work continues to date e.g. Čermák and Nechvátal (2022). Fractional and integrodifferential equations of pantograph type have also been studied (Lachouri et al. 2023).

There has been a lot of activity in extending results to stochastically perturbed pantograph equations and equations with unbounded delay, especially in the direction of asymptotic behaviour and numerical methods (see e.g.Appleby 2005; Baker and Buckwar 2000, 2005; Fan et al. 2007; Huang et al. 2012; Hu et al. 2012; Jiang et al. 2010; Liu and Chen 2008; Meng et al. 2011; Meng and Yin 2012; Pavlovic and Jankovic 2012; Yan and Huang 2013; Yang et al. 2010). The first such work on asymptotic behaviour is Appleby and Buckwar (2016). In this field of stochastic asymptotic analysis, papers concerning power–law behaviour and numerical methods (Wu et al. 2022; Guo and Li 2019; Milosevic 2014), large deviations (Siva Ranjani et al. 2023), and neutral equations (Caraballo et al. 2021) and Lévy noise (Mao et al. 2020) have also been recently considered. Many *sufficient conditions* have been obtained with respect to which solutions of stochastic equations exhibit convergence to a limit, either in pth moment or pathwise. However, what appears to be absent in this literature are results which *characterise* certain types of asymptotic behaviour in terms of forcing terms, such as, for instance, convergence to the equilibrium of the unforced equation, or asymptotic rates of growth or decay of forced solutions; it is one of the chief contributions of this paper to supply such asymptotic characterisations. But as we indicate now, we believe that the techniques in this paper have much wider application than to pantograph equations.

In this paper, we consider perturbations of (1) where the intensity of the forcing terms is state–independent. Both deterministic and stochastic equations are considered, and in the case of stochastic equations, we allow the solution to be forced both deterministically and stochastically. The main contribution of the work is to give a *characterisation* of certain convergence, boundedness, and growth estimates in terms of the forcing terms. By a characterisation of an asymptotic property **A** of solutions, we mean that solutions possess property **A** holds if and only if some conditions **C** are satisfied by the forcing terms; moreover, the conditions **C** should be expressible without recourse to properties of the solution of the forced equation, or other auxiliary objects that cannot be directly computed from the problem data. Ideally, such conditions should be easy to check. The characterisations we give satisfy these natural requirements.

The techniques used in this paper are especially suited to considering such "additive" rather than "multiplicative" perturbations (the terminology and analysis is more commonplace for stochastic equations, but is appropriate for deterministic equations too). Stochastic equations with such state–dependent noise have been studied in the papers listed above. Our analysis of additive noise equations relies on a decomposition approach to solutions of stochastic delay differential equations, which allows such solutions to be built from solutions of stochastic ordinary differential equations with non–differentiable paths, and solutions of random functional differential equations which are continuously differentiable. A similar approach can be applied to equations with multiplicative noise, and a sketch of possible research in that direction is given in Sect. 6.

Other generalisations, to scalar equations with a general unbounded delay, or to multidimensional equations, are considered in Sect. 6, and sketch proofs are given which show that at least basic convergence and boundedness results still hold, giving reassurance that the theory is not confined to the narrow class of scalar, proportional delay equations. However, the main thrust of our paper concerns pantograph equations, partly due to their celebrated role in the historical development of the theory of unbounded delay equations, and partly because, since the asymptotic behaviour of the unperturbed pantograph equation is essentially perfectly understood (in contrast to many other unbounded delay equations) we can exploit this knowledge to obtain equally pristine results in the perturbed case, thereby demonstrating the sharpness of our approach.

In doing so, we are able to improve decay estimates on perturbed equations, show that these estimates are optimal in certain cases, accurately characterising when perturbation sizes cause solutions to transition from preserving the asymptotic behaviour of the unperturbed equation to slower rates of decay (or even growth) when the perturbation are larger than a critical size. In fact, by proving both big-O and little-o growth estimates on solutions throughout, we are able to characterise in many cases (in terms of an easily estimated functional of the perturbation, which is entirely independent of the structure of the unperturbed equation) when the solution of the perturbed equation has a given order of growth (or decay). The combination of (a) sharp estimation of growth rates (b) corresponding correct determination of the transition point from the small to the large perturbation regime, and (c) characterisation of the growth conditions on the forcing terms which give rise to certain rates of growth or decay in solutions, are to the best of our knowledge new in the asymptotic study of unbounded delay differential equations.

Another novel feature in our analysis is to be able, for functional differential equations which are simultaneously deterministically and stochastically perturbed, to characterise certain important asymptotic behaviours. In particular, we are able to characterise exactly when solutions of perturbed equations are convergent or bounded, and the characterising conditions are very easy to check. Such a characterisation, with only continuity side conditions on forcing terms being required, seems new to the literature, *even for non–delay stochastic differential equations*. Better yet, we sketch at the end of Sect. 5 an argument which suggests strongly that *such a characterisation might well apply to a very broad class of functional differential equations* which can have finite or unbounded delay, and which does not possess any of the special structure of the pantograph equations which we study here.

We note that a great deal of the sharpness achieved in our results comes from a very careful asymptotic characterisation of solutions of scalar linear *ordinary* and stochastic differential equations. Remarkably, given the simplicity of the equations involved, such a characterisation does not seem to be present in the literature, and part of the purpose of this paper is to show that a thorough *development of this theory for linear ordinary differential equations will unlock a characterising perturbation theory for a very wide class of functional differential equations*.

We have chosen perturbed pantograph equations to do this, because the perturbation theory of non–autonomous delay differential equations (of which the pantograph equation is a well–studied example) has traditionally found comparatively little use for resolvents or fundamental solutions, at least compared with autonomous equations. Indeed, the analysis in this paper likewise makes no use whatsoever of resolvents or variation of constants formulae. In itself, this is neither noteworthy nor surprising. However, in other works, the authors have given asymptotic characterisation results for perturbations of autonomous deterministic and stochastic functional and Volterra equations, in which deterministic and stochastic variation of constants formulae are in constant use. This might leave one to believe that such formulae play a critical role in making the perturbation theory both achievable and sharp. *However, this paper shows that while these resolvent–based representations are certainly useful when available, they do not appear to be essential for the development of a sharp perturbation theory*. As a result, the pantograph equation represents an interesting test case for a sharp perturbation theory that applies to other non–autonomous or even non–linear functional differential equations.

1.1 Technical Introduction

To make some of this general discussion precise, we let f and σ be continuous deterministic functions, B be a standard Brownian motion, and a and b be real numbers obeying

$$b < 0, \quad |b| > |a|, \tag{2}$$

and consider

$$x'(t) = ax(qt) + bx(t) + f(t),$$
$$dX(t) = (aX(qt) + bX(t) + f(t))\,dt + \sigma(t)\,dB(t).$$

Equation (2) is a necessary and sufficient condition for all solutions of (1) to tend to zero as $t \to \infty$ (Kato and McLeod 1971). Throughout we take $a \neq 0$, since this case deals with non–delayed equations. In the case of the deterministic equation, we give necessary and sufficient conditions on f for which solutions will tend to zero, or for which solutions are bounded. We also characterise when solutions x are unbounded with a certain order of growth, or decay to zero more slowly than those of (1), and give sharp conditions under which the decay rate of x is the same as that of (1). For the stochastic equation, we give conditions on f and σ which characterise when $X(t) \to 0$ as $t \to \infty$ a.s. and when X is almost surely bounded. Results for multi–dimensional equations are also sketched. In later works, we will develop quantitative pathwise estimates for X, and relax the second inequality in (2), so as to study the situation when the underlying unforced Eq. (1) has unbounded solutions.

In short, therefore this paper shows that provided all solutions of (1) tend to zero, we can find coincident necessary and sufficient conditions on the forcing terms f and σ (which do not depend on the parameters b, a and q) under which the solutions are convergent to zero; in particular, it is not necessary to strengthen conditions on the underlying equation in order to preserve stability, merely to impose appropriate smallness conditions on f and σ.

The results in this paper develop work of the authors for certain types of *uniformly asymptotically stable* and *autonomous* linear differential and Volterra equations which are forced both deterministically and stochastically by time–independent forcing terms. The nature of conditions we impose on f and σ for pantograph equations are identical to those needed in the autonomous case, and we avail of existing results from Appleby and Lawless (2024c) for ordinary differential equations in our proofs. Related results for the mean square asymptotic behaviour of affine equations, where there is also state–dependence in the diffusion coefficient, are given in Appleby and Lawless (2024a). For the stochastic differential equation $dY_0(t) = -Y_0(t)\,dt + \sigma(t)\,dB(t)$ it is shown in Appleby et al. (2011) that $Y_0(t) \to 0$ as $t \to \infty$ a.s. if and only if $S(\epsilon) < +\infty$ for all $\epsilon > 0$, where

$$S(\epsilon) := \sum_{n=1}^{\infty} \sqrt{\int_{n-1}^{n} \sigma^2(s)\,ds}\, \exp\left(-\epsilon \int_{n-1}^{n} \sigma^2(s)\,ds\right). \tag{3}$$

In Appleby and Lawless (2024c) (as discussed in Sect. 4), it is shown that solutions of the ordinary differential equation $y'(t) = -y(t) + f(t)$ obey $y(t) = O(\gamma(t))$ as $t \to \infty$ for non–decreasing and so–called subexponential weight functions γ if and only if f_θ, defined by $f_\theta(t) := \int_{(t-\theta)^+}^{t} f(s)\,ds$ for $(t,\theta) \in [0,\infty) \times [0,1]$, obeys

$$\sup_{0 \le \theta \le 1} |f_\theta(t)| \le K\gamma(t), \quad t \ge 0, \quad \text{for some } K > 0. \tag{4}$$

(In the above, and throughout the paper, we make extensive use of the standard "big O" and "little o" Landau notation, as well as the notation $a(t) \sim b(t)$, $t \to \infty$ for real–valued functions a and b to signify that $a(t)/b(t) \to 1$ as $t \to \infty$). Finally, in this volume, we also show that the asymptotic behaviour of f_θ characterises the situation when solutions of deterministic functional and Volterra differential equations have time averages (see Appleby and Lawless 2024b).

Therefore, we see that the asymptotic behaviour of f_θ for $\theta \in [0, 1]$ and $\sigma_1^2(t) := \int_{(t-1)^+}^t \sigma^2(s)\, ds$ appear to characterise many interesting types of asymptotic behaviour of solutions of forced autonomous equations. In this work, we show that this is true for perturbed pantograph equations also. *Therefore, the evidence accumulates in favour of the overall viewpoint that situations in which important qualitative and quantitative asymptotic properties of solutions of perturbed differential systems (whether deterministic or stochastic, memory–dependent or not, autonomous or not, and perhaps even linear or not), may be characterised by in terms of perturbations through* $\{f_\theta : \theta \in [0, 1]\}$ *and* σ_1^2.

We recall some facts about the asymptotic behaviour of solutions of (1); all results appear in Kato and McLeod (1971). There is no loss of generality in taking the initial condition to be unity; denoting the solution of the differential equation with initial condition $\xi \in \mathbb{R}$ by $u(\cdot; \xi)$, we have that $u(t; \xi) = u(t; 1)\xi$ for all $t \geq 0$. Notice that if $\xi = 0$, then $u(t; \xi) = 0$ for all $t \geq 0$, so that the zero function is an equilibrium solution for all choices of the parameters (when $a + b = 0$, every constant is a solution, but we do not study this phenomenon).

If $b > 0$, $u(t)/e^{bt}$ tends to a finite limit, so solutions exhibit real exponential growth: we will not study this case here. When $b < 0$, however, power–law asymptotic behaviour occurs. Specifically, since $a \neq 0$ and $b < 0$ we may define

$$\kappa := -\frac{\log |b/a|}{\log(1/q)}, \tag{5}$$

and $u(t) = O(t^\kappa)$ as $t \to \infty$. Indeed, there are no non–trivial solutions that are $o(t^\kappa)$ as $t \to \infty$, so this estimate is very sharp (Kato and McLeod 1971, Theorem 3). More precise information regarding oscillation of solutions are known, but since we focus on sharp upper estimates of solutions of perturbed equations, we do not seek such refined results.

This discussion explains in part why we wish to confine attention in this paper to the case when (2) holds, but does not mean that it is not interesting, or impossible, to give a perturbation theory in the other parameter regimes mentioned above, and it is a task that we will return to in other works.

2 A Fundamental Estimate and Large Perturbation Estimates

We study the deterministic pantograph equation first, partly for reasons of simplicity, but also because solutions of the stochastic equation can be written in terms of the deterministic equation, when viewed path–by–path in the sample space. Therefore the analysis of the deterministic equation is fundamental. Suppose that

$$f \in C([0, \infty); \mathbb{R}), \tag{6}$$

and consider the forced pantograph equation

$$x'(t) = ax(qt) + bx(t) + f(t), \quad t \geq 0; \quad x(0) = \zeta \in \mathbb{R}. \tag{7}$$

There is a unique continuous solution to this initial value problem (see e.g. Appleby and Buckwar 2010). If (2) holds, then $u(t) \to 0$ as $t \to \infty$. It is therefore reasonable to ask under what "smallness" conditions on the forcing function f certain asymptotic properties (such as convergence to zero, or rates of decay) of u are preserved; on the other hand, we may be interested in understanding the properties of solutions when f does not obey smallness properties which preserve properties of u.

To this end, in this section, we establish a "fundamental" estimate on the size of solutions of perturbed pantograph equations which enables us to obtain not only convergence and stability results, but also (after further analysis later in the paper) very sharp asymptotic bounds, especially when perturbations are, in a certain sense, small. The results are of a "sufficiency" type, which is to say, if sufficiently strong asymptotic conditions are imposed on the forcing term, this will generate a rate of decay on the solution. A full characterisation of asymptotic behaviour of solutions will require "necessity" results, as well as some transformations and decomposition results; this is because our fundamental lemma is usually applied to a perturbed pantograph equation which results from such a decomposition, rather than to the original pantograph equation itself. We will also need estimates which deal with large perturbations, and these will be proved separately. Therefore, our fundamental estimate, although a key tool in the asymptotic analysis, provides only part of the picture.

This approach differs from our analysis of autonomous differential and Volterra integrodifferential equations, where we exploit variation of constants formulae and known (integrable) behaviour of the differential resolvent. Although perturbed pantograph equations admit such variation of constants formulae, the non–autonomous character of the equation makes it difficult to get good asymptotic information about the underlying resolvent, and this motivates an approach avoiding this problem. In this sense, our approach bears relation to that taken in Appleby and Buckwar (2010).

In order to state our fundamental estimate, we introduce some notation. For $a \neq 0$, there are (in general, complex valued) solutions k to the equation $aq^k + b = 0$. If $k_0 \in \mathbb{C}$ is such a solution, $\kappa := \text{Re}(k_0)$ is uniquely defined, and given by (5). Let

$$c := \log q < 0, \tag{8}$$

$$\alpha := |a/b|, \tag{9}$$

so that $e^c = q$, and under (2), $\alpha \in (0, 1)$.

Lemma 1 *Suppose $q \in (0,1)$ and (2) holds. Let $\varphi \in C([0,\infty);\mathbb{R})$ and z be the unique continuous solution to*

$$z'(t) = az(qt) + bz(t) + \varphi(t), \quad t \geq 0; \quad z(0) = \xi \in \mathbb{R}. \tag{10}$$

Let α, c be defined by (9), (8) *and define $I_n := [s_0 - nc, s_0 - (n+1)c]$. Let*

$$K_n := \sup_{s \in I_n} e^{-\kappa s}|z(e^s)|, \tag{11}$$

and let $\epsilon_n > 0$ be a sequence such that

$$\sup_{s \in I_{n+1}} |\varphi(e^s)| \leq \epsilon_{n+1}. \tag{12}$$

(i) If (ϵ_n/α^n) is summable, $K_n \leq K^*$, and $z(t) = O(t^\kappa)$ as $t \to \infty$.
(ii) If (ϵ_n/α^n) is not summable, then $K_n \leq K^* \sum_{j=1}^{n} \frac{\epsilon_j}{\alpha^j}$ and

$$\sup_{t \in [e^{s_0 - nc}, e^{s_0 - nc}e^{-c}]} |z(t)| \leq K^* \alpha^n \sum_{j=1}^{n} \frac{\epsilon_j}{\alpha^j}. \tag{13}$$

The proof of the result is inspired especially by that of Kato and McLeod (1971, Theorem 3(i)) and also of Čermák (2002, Lemma 3.3). It differs from the former in considering perturbed equations. Concerning the latter paper and result, we note that it considers more general perturbed delay equations, and yields estimates on their solutions using *pointwise* conditions on forcing terms. However, the summability conditions in Lemma 1 hint to the development of *integrability* conditions on forcing terms in our analysis, which will ultimately yield extra sharpness to our results. In addition, our decomposition approach to solutions of the perturbed pantograph equations, shown in the next section, has the effect of turning pointwise conditions on the forcing terms into average conditions too, which will further sharpen our deterministic results. The decomposition approach will even enable us exploit the fundamental lemma for stochastic equations, where (poor) pointwise asymptotic behaviour and regularity (i.e. non–differentiability) of the forcing terms would confound the analysis in Čermák (2002).

At this instant, the reader must take these claims on trust, since it is not immediately apparent how sharp or useful the estimate in part (ii) is, nor that it can be applied to study stochastic equations. In fact, quite a lot of further analysis is needed so that full value can be extracted from this lemma. Nevertheless, we will eventually show that this fundamental lemma (a) develops essentially unimprovable size estimates on the size of solutions of deterministic equations, which improve all existing asymptotic estimates; (b) allows us to give characterisations of the almost sure convergence and boundedness of stochastically perturbed equations, results hitherto absent from the literature.

We turn to these matters later. However, we can immediately prove two direct corollaries of Lemma 1, under the assumption (2), giving some immediate reassurance that the Lemma 1 will be of some utility.

Theorem 1 *Suppose* $q \in (0, 1)$ *and* (2) *holds. Let* $\varphi \in C([0, \infty); \mathbb{R})$ *and* z *be the unique continuous solution to* (10).

(a) If $\varphi(t) \to 0$ *as* $t \to \infty$, *then* $z(t) \to 0$ *as* $t \to \infty$.
(b) If φ *is bounded, then* z *is bounded.*

Proof If $\varphi(t) \to 0$ as $t \to \infty$, we may take equality in (12), and note the hypothesis forces $\epsilon_n \to 0$ as $n \to \infty$. By (2), $\alpha \in (0, 1)$. Either $\lambda_n = \epsilon_n/\alpha^n$ is summable or not.

Suppose λ_n is summable. Then, by Lemma 1 (i), $\sup_{t\in[e^{s_0-nc}, e^{s_0-nc}e^{-c}]} |z(t)| \le K^*[e^{(s_0-nc)}]^\kappa$, so $z(t) \to 0$ as $t \to \infty$, because $\kappa < 0$ and $c < 0$.

Suppose λ_n is not summable. Since $(1/\alpha^n)$ is divergent, by Toeplitz lemma (see Shiryaev 1996, p. 390), $\sum_{j=1}^n \epsilon_j \alpha^{-j} = o\left(\sum_{j=1}^n \alpha^{-j}\right)$ as $n \to \infty$, since $\epsilon_n \to 0$ as $n \to \infty$. On the other hand $\sum_{j=1}^n \alpha^{-j} \sim \alpha^{-n}(1 - \alpha^{-1})$ as $n \to \infty$, so $\alpha^n \sum_{j=1}^n \epsilon_j \alpha^{-j} \to 0$ as $n \to \infty$. Since the righthand side of (13) tends to 0 as $n \to \infty$, $z(t) \to 0$ as $t \to \infty$.

In the case when φ is bounded on $[0, \infty)]$ by $\overline{\varphi}$, say, we may fix $\epsilon_n = \overline{\varphi}$. By part (i) of the lemma, which holds whether (ϵ_n/α^n) is summable or not, we have that $K_n \le K^*\overline{\varphi} \sum_{j=1}^n \alpha^{-j} \le \bar{K}\alpha^{-n}$ since $\alpha < 1$, for some $\bar{K} > 0$. Thus $K_n\alpha^n \le \bar{K}$. Revisiting the above estimates gives

$$\sup_{t\in[e^{s_0-nc}, e^{s_0-nc}e^{-c}]} |z(t)| \le K_n\alpha^n e^{\kappa s_0} \le \bar{K}e^{\kappa s_0} =: \bar{z},$$

and so z is bounded, as required. □

As we see in Sect. 6, this qualitative result generalises to equations with unbounded delay, using methods from Appleby and Buckwar (2010) (see also Theorem 4 below). But our primary interest in the fundamental Lemma derives from quantitative estimates that it generates, stated below. For these estimates, we recall that a (measurable) function $\gamma : [0, \infty) \to (0, \infty)$ is regularly varying at infinity with index α if $\lim_{t\to\infty} \gamma(\lambda t)/\gamma(t) = \lambda^\alpha$ for all $\lambda > 0$. The notation $\gamma \in \mathrm{RV}_\infty(\alpha)$ is used. We will exploit properties of regularly varying functions throughout this work; see Bingham et al. (1991), especially Chap. 1, for relevant results.

Theorem 2 *Suppose that* $q \in (0, 1)$ *and* (2) *holds. Let* $\varphi \in C([0, \infty); \mathbb{R})$ *and* z *be the unique continuous solution to* (10), *Let* $\varphi(t) = O(\gamma(t))$ *as* $t \to \infty$, *where* γ *is regularly varying at infinity. Then*

$$z(t) = O\left(t^\kappa \int_1^t \frac{\gamma(s)}{s^{\kappa+1}}\, ds\right), \quad t \to \infty.$$

The proof is deferred to the end. Since the theorem gives only bounds on solutions, it is reasonable to ask whether (i) the bounds are sharp; (ii) the transition from the rate of decay of the solutions of (1) to slower decay rates, do indeed occur when $t \mapsto \varphi(t)/t^{\kappa+1}$ changes from being an integrable to a non–integrable function. The following example shows that Theorem 2 is sharp in both regards. Furthermore, in Sect. 4, we give another example which shows that the "big O" estimate cannot be improved in general to results of the form $z(t) \sim ct^{\kappa}$ as $t \to \infty$, at least in the case of small perturbations (i.e. when $t \mapsto \gamma(t)/t^{\kappa+1}$ is integrable).

Theorem 3 *Let $a > 0$, $b < 0$, $\varphi \in C(\mathbb{R}^+)$ and z the continuous solution to* (10).

(a) For every $C > 0$ and every $\gamma \in RV_{\infty}(\kappa)$ such that $t \mapsto \gamma(t)/t^{\kappa+1}$ is not in $L^1(0,\infty)$, there exists $\varphi(t) \sim C\gamma(t)$ such that for some $D > 0$,

$$z(t) \sim Dt^{\kappa} \int_1^t \frac{\gamma(s)}{s^{\kappa+1}}\,ds, \quad t \to \infty.$$

(b) For every $C > 0$ and every $\gamma \in RV_{\infty}(\kappa)$ such that $t \mapsto \gamma(t)/t^{\kappa+1}$ is in $L^1(0,\infty)$, there exists $\varphi(t) \sim C\gamma(t)$ such that $z(t) \sim Dt^{\kappa}$ as $t \to \infty$ for some $D > 0$.

Proof For part (a), write $\psi(t) := \gamma(t)/t^{\kappa+1}$. Then $\psi \in \mathrm{RV}_{\infty}(-1)$ and ψ is not integrable. We may take ψ to be a smooth regularly varying function. Let $\Psi(t) := \int_{1/q}^t \psi(s)\,ds$. Then Ψ is regularly varying with index zero, and $\Psi(t) \to \infty$ as $t \to \infty$. Moreover, Ψ is smoothly regularly varying, with $t\Psi'(t)/\Psi(t) \to 0$ as $t \to \infty$. Set $z(t) := Dt^{\kappa}\Psi(t)\, t \geq q$ and $\varphi(t) := z'(t) - bz(t) - az(qt)$, $t \geq 1$. Then z satisfies (7) on $[1,\infty)$. Moreover $z'(t) = Dt^{\kappa-1}\Psi(t)(\kappa + t\Psi'(t)/\Psi(t)) \sim \kappa Dt^{\kappa-1}\Psi(t)$ as $t \to \infty$. For $t \geq 1$, using the fact that $b + aq^{\kappa} = 0$, we have that $bz(t) + az(qt) = bDt^{\kappa}(\Psi(t) - \Psi(qt))$. Compare the asymptotic behaviour of z' and the last term, to get the dominant asymptotic behaviour in φ. $\Psi' = \psi$ is regularly varying with index -1, the uniform convergence theorem for regularly varying functions gives

$$\frac{\Psi(t) - \Psi(qt)}{t\Psi'(t)} = \int_q^1 \frac{\Psi'(\lambda t)}{\Psi'(t)}\,d\lambda \to \int_q^1 \lambda^{-1}\,d\lambda = \log(1/q), \quad t \to \infty.$$

Hence $bz(t) + az(qt) \sim b\log(1/q)Dt^{\kappa+1}\psi(t)$ as $t \to \infty$, so this expression is in $RV_{\infty}(\kappa)$, while z' is in $RV_{\infty}(\kappa - 1)$. Therefore $z'(t) = o(bz(t) + az(qt))$, and so we have $\varphi(t) \sim -(bz(t) + az(qt)) \sim Ct^{\kappa+1}\psi(t)$, where $C = -b\log(1/q)D > 0$.

For part (b), let ψ be a smooth regularly varying function with index -1. Then Ψ is regularly varying with index zero, but now since ψ is integrable $\Psi(t) \to L \in (0,\infty)$ as $t \to \infty$. Moreover, Ψ is also smoothly regularly varying, and so $t\Psi'(t)/\Psi(t) \to 0$ as $t \to \infty$. Next let $D > 0$ be arbitrary and set

$$z(t) = \frac{D}{L+1}\left(t^{\kappa} + t^{\kappa}\int_0^t \psi(s)\,ds\right), \quad t \geq q.$$

Set $\varphi(t) = z'(t) - (bz(t) + az(qt))$ for $t \geq 1$. Then z is the unique continuous solution of (7) on $[1, \infty)$ and $z(t) \sim Dt^\kappa$ as $t \to \infty$. We determine the asymptotic behaviour of f. Write $\tilde{D} = D/(L+1)$. To start, note $z'(t) = \tilde{D}\kappa t^{\kappa-1}(\kappa + \kappa\Psi(t) + t\psi(t))$. Since $t\psi(t) = t\Psi'(t)/\Psi(t) \cdot \Psi(t) \to 0 \cdot L{=}0$ as $t \to \infty$, $z'(t)$ is order $t^{\kappa-1}$ as $t \to \infty$. Next

$$bz(t) + az(qt) = b\tilde{D}t^\kappa \int_{qt}^{t} \psi(s)\,ds,$$

so, proceeding as before we get $bz(t) + az(qt) \sim b\tilde{D}t^{\kappa+1}\psi(t)\log(1/q)$ as $t \to \infty$, which is regularly varying with index κ, and dominates $z'(t)$. Letting $C := -bD(L+1)^{-1}\log(1/q) > 0$, gives $\varphi(t) \sim Ct^{\kappa+1}\psi(t) = C\gamma(t)$ as $t \to \infty$, as needed. □

Perturbations with decay faster than a power law are covered by Theorem 2: if $\varphi(t) = o(t^{-M})$ for all $M > 0$, then $z(t) = O(t^\kappa)$, by taking $\gamma(t) = t^{\kappa-2}$.

For larger perturbations that may not have a power–law type form, the following bound can be established directly, without recourse to the fundamental lemma.

Theorem 4 *Suppose that $q \in (0, 1)$ and (2) holds. Let $\varphi \in C([0, \infty); \mathbb{R})$ and z be the unique continuous solution to (10). Let γ be C^1 and non–decreasing.*

(A) If $\varphi(t) = O(\gamma(t))$ as $t \to \infty$, then $z(t) = O(\gamma(t))$, as $t \to \infty$.
(B) If $\varphi(t) = o(\gamma(t))$ and $\gamma(t) \to \infty$ as $t \to \infty$, then $z(t) = o(\gamma(t))$, as $t \to \infty$.

Proof The proof is inspired by those in Appleby and Buckwar (2010, 2016). Denote by $D_+u(t) = \limsup_{h\to 0^+}(u(t+h) - u(t))/h$ the Dini derivative of any continuous function u. Since $b < 0$, by considering the cases where $z(t)$ is positive, negative or zero, we arrive at the consolidated estimate $D_+|z(t)| \leq b|z(t)| + |a||z(qt)| + |\varphi(t)|$ for $t \geq 0$. We note that if z_+ is any positive $C^1(0, \infty)$ function satisfying $z_+(0) > |z(0)|$ and $z_+'(t) > bz+(t) + |a|z_+(qt) + |\varphi(t)|$ for all $t \geq 0$, then $z_+(t) > |z(t)|$ for all $t \geq 0$, by standard comparison arguments. By hypothesis, there exists $C > 0$ such that $|\varphi(t)| \leq C\gamma(t)$ for $t \geq 0$. Take $K > \max(C/(|b| - |a|), |z(0)|/\gamma(0))$, and set $z_+(t) = K\gamma(t)$ for $t \geq 0$. Then $z_+(0) = K\gamma(0) > |z(0)|$ and for $t \geq 0$ we have

$$\begin{aligned} bz_+(t) + |a|z_+(qt) + |\varphi(t)| &\leq bK\gamma(t) + |a|K\gamma(qt) + C\gamma(t) \\ &\leq (bK + |a|K + C)\gamma(t) < 0 < K\gamma'(t) = z_+'(t), \end{aligned}$$

using the monotonicity of γ. Thus $|z(t)| < z_+(t) = K\gamma(t)$ for all $t \geq 0$, as needed.

For part (B), by hypothesis, we have $|\varphi(t)| < \epsilon\gamma(t)$, $t \geq T_1(\epsilon)$. For $t \geq qT_1(\epsilon)$, define $z_\epsilon(t) = \epsilon\gamma(t)/(|b| - |a|) + C(\epsilon)$ where $C(\epsilon) := 1 + \max_{t\in[qT_1,T_1]}|z(s)| > 0$. Then $z_\epsilon(t) > |z(t)|$ for $t \in [qT_1, T_1]$. For $t \geq T_1$ we have

$$\begin{aligned} bz_\epsilon(t) + |a|z_\epsilon(qt) + |\varphi(t)| &= (b+|a|)C + \epsilon\left(\frac{b}{|b|-|a|}\gamma(t) + \frac{|a|}{|b|-|a|}\gamma(qt)\right) + |\varphi(t)| \\ &\le (b+|a|)C + \epsilon\frac{b+|a|}{|b|-|a|}\gamma(t) + \epsilon\gamma(t) \\ &= (b+|a|)C < 0 < \frac{1}{|b|-|a|}\epsilon\gamma'(t) = z_\epsilon'(t). \end{aligned}$$

Thus $|z(t)| < z_\epsilon(t) = \epsilon\gamma(t)/(|b|-|a|) + C(\epsilon)$ for $t \ge T_1(\epsilon)$. Dividing by $\gamma(t)$ on both sides, letting $t \to \infty$, and then $\epsilon \to 0$, we get the desired result. □

It is easy to show that these bounds are sharp in at least some cases by considering an example. Suppose that $\gamma'(t)/\gamma(t) \to 0$ as $t \to \infty$ and $t\gamma'(t)/\gamma(t) \to \infty$ as $t \to \infty$. Then γ is increasing subexponentially (by the first limit, see (18) and the discussion thereafter), but faster than any power, by the second. Thus $\gamma(qt)/\gamma(t) \to 0$ as $t \to \infty$. Let $z(t) = D\gamma(t)$ for $t \ge 0$, and define $\varphi(t) = D[\gamma'(t) - b\gamma(t) - a\gamma(qt)]$ for $t \ge 0$. Then z obeys (10), $\varphi(t) \sim -bD\gamma(t)$ as $t \to \infty$, and $z(t) \sim D\gamma(t)$ as $t \to \infty$, confirming the sharpness in part (a).

The "little–o" result in part (B) of Theorem 4 is less traditional, and the reader may query its value. However, we will later show that such results can be used in conjunction with the conventional "big–O" results to characterise when solutions obey $\limsup_{t\to\infty} |z(t)|/\gamma(t) \in (0,\infty)$, thereby establishing the exact asymptotic order of the running maxima of solutions.

We note that Theorem 4 *does not* give results where an exact rate of growth of the solution is given e.g., $z(t) \sim c\gamma(t)$ as $t \to \infty$, where $\gamma(t) \to \infty$ as $t \to \infty$. To give conditions on the forcing terms under which this would occur is also of interest, and will form part of a future investigation.

3 Deterministically Forced Equation: First Results, Auxiliary ODEs

Theorem 1 shows that $f(t) \to 0$ as $t \to \infty$ is a *sufficient* condition in order for the solution x of (7) to obey $x(t) \to 0$ as $t \to \infty$. With a little extra work, it can be used to develop a necessary and sufficient condition. Consider

$$y'(t) = -y(t) + f(t), \quad t \ge 0; \quad y(0) = 0. \tag{14}$$

Equation (14) has a unique continuous solution, given by $y(t) = \int_0^t e^{-(t-s)} f(s)\,ds$ for $t \ge 0$.

We now show that the asymptotic behaviour of (7) is determined solely by y.

Theorem 5 *Suppose that $q \in (0,1)$ and (2) holds. Let $f \in C([0,\infty);\mathbb{R})$ and x be the unique continuous solution to (7). Then the following are equivalent*

(A) $y(t) \to 0$ *as* $t \to \infty$*;*
(B) For every $\theta > 0$, $\lim_{t\to\infty} \int_t^{t+\theta} f(s)\,ds = 0$*;*
(C) $x(t) \to 0$ *as* $t \to \infty$.

Proof The equivalence of (A) and (B) is known; we show (A) and (C) are equivalent. Suppose (A) holds. Set $z(t) = x(t) - y(t)$ for $t \ge 0$, $z(0) = x(0) = \zeta$. Define

$$\varphi(t) = ay(qt) + (b+1)y(t), \quad t \ge 0. \tag{15}$$

Note that φ is continuous. Since $z' = x' - y'$, we have that z obeys (10). If (A) holds, $\varphi(t) \to 0$ as $t \to \infty$. Since z obeys (10), we may apply Theorem 1 part (a) to it, so $z(t) \to 0$ as $t \to \infty$. But $y(t) \to 0$ as $t \to \infty$ and $x = y + z$, so (C) follows.

To show (C) implies (A), re–write (7) according to $x'(t) = -x(t) + \{x(t) + bx(t) + ax(qt)\} + f(t)$ for $t \ge 0$. Let $g(t)$ be the sum of the last two terms; then $x' = -x + g$ and thus integrating by parts and rearranging we get

$$y(t) = x(t) - \xi e^{-t} - \int_0^t e^{-(t-s)}\{(1+b)x(s) + ax(qs)\}\,ds, \quad t \ge 0. \tag{16}$$

Since (C) holds, the first two terms on the righthand side of (16) tend to zero as $t \to \infty$. By (C) the term in curly brackets in the integrand tends to zero as $s \to \infty$. The integral in (16) is the convolution of an $L^1(0,\infty)$ function and a function in $BC_0(0,\infty)$, so it tends to 0 as $t \to \infty$. Thus $y(t) \to 0$ as $t \to \infty$, proving (A). □

We note that this appears to be the first instance in the literature of a result which *characterises* conditions on perturbations which preserve the asymptotic stability of non–autonomous functional differential equations. The same remark can be made concerning boundedness; see Theorem 7 later. Sufficient conditions for which convergence to the equilibrium are preserved, which already appear in the literature, include $f(t) \to 0$ as $t \to \infty$ or $f \in L^1(0,\infty)$ (we note that both imply $y(t) \to 0$ as $t \to \infty$), but Theorem 5 shows that neither are necessary.

This leads us to ask what functions f allow $y(t) \to 0$ as $t \to \infty$ which do not satisfy the familiar stability conditions above. This matter has been explored somewhat in Appleby and Lawless (2024a). Two classes of functions with bad "pointwise" behaviour, but good "average" behaviour naturally emerge, but we do not claim that these characterise all the functions with stability preserving properties.

The first class are functions which exhibit high frequency oscillation (even with unbounded amplitude, provided the frequency grows faster than the amplitude). A prototype for such a function is

$$f(t) = e^{\beta t}\sin(e^{\theta t}), \quad t \ge 0,$$

where $\theta > \beta > 0$. This function, by most measures, is ill–behaved as $t \to \infty$, oscillating between exponentially growing upper and lower envelopes, with increasingly higher frequency. Obviously, $f(t) = O(e^{\beta t})$ as $t \to \infty$, yet $y(t) = O(\max(e^{-t}, e^{-(\theta-\beta)t}))$ as $t \to \infty$. Therefore, as $y(t) \to 0$ as $t \to \infty$, we have that $x(t) \to 0$ as $t \to \infty$.

Another class of functions f which give rise to convergence of y, and which do not satisfy $f(t) \to 0$ as $t \to \infty$, or $f \in L^1(\mathbb{R}^+)$, are nonnegative functions with unboundedly large "spikes" of very short duration. Notice that if f is a non–negative function, the condition (B) in Theorem 5 is equivalent to

$$\int_n^{n+1} f(s)\,ds \to 0, \quad n \to \infty,$$

where n tends to infinity through the integers. We construct such a function now; let let $n \in \mathbb{N}$, and suppose $0 < w_n < 1$ and $w_n \to 0$ as $n \to \infty$. Let f be zero on $[n, n+1]$ except on the interval $[n + 1/2 - w_n/2, n + 1/2 + w_n/2]$. Let f be linear on $[n + 1/2 - w_n/2, n + 1/2]$, passing through the points $(n + 1/2 - w_n/2, 0)$ and $(n + 1/2, h_n)$ where $h_n > 0$ and $h_n \to \infty$ as $n \to \infty$, and likewise let f be linear on $[n + 1/2, n + 1/2 + w_n/2]$, passing through the points $(n + 1/2, h_n)$ and $(n + 1/2 + w_n/2, 0)$. Then f is continuous and nonnegative and obeys

$$\int_n^{n+1} f(s)\,ds = \frac{1}{2} w_n h_n.$$

Clearly, $\max_{n \le s \le n+1} f(s) = h_n$, so if we arrange that $h_n \to \infty$ as $n \to \infty$, while $w_n h_n \to 0$ as $n \to \infty$, we have a function which satisfies condition (B), and obeys $\limsup_{t\to\infty} f(t) = +\infty$. Indeed, since h_n can be chosen to grow arbitrarily quickly, the running maximum of f can grow arbitrarily quickly in time. It is also easy to arrange that f is not in $L^1(\mathbb{R}^+)$, since for any $N \in \mathbb{N}$ we have

$$\int_0^{N+1} |f(s)|\,ds = \frac{1}{2} \sum_{n=0}^{N} w_n h_n,$$

so it is merely necessary to arrange for $w_n h_n$ to be divergent. Thus, we may have f having partial maxima which grow arbitrarily fast (let us take $h_n \to \infty$ as $n \to \infty$ in any way we choose such that $h_n > 1$ and h increasing); picking $w_n = \frac{1}{n h_n}$, we have that $0 < w_n < 1$, $w_n \to 0$ as $n \to \infty$, condition (B) holds, but $\limsup_{t\to\infty} f(t) = +\infty$ and $f \notin L^1(\mathbb{R}^+)$. Thus, for such f, we have that $y(t) \to 0$ as $t \to \infty$, and hence $x(t) \to 0$ as $t \to \infty$.

Theorem 5 motivates a characterisation of the asymptotic behaviour of the linear differential equation (14) given in Appleby and Lawless (2024c), from which we will presently draw. The conditions and arguments in Appleby and Lawless (2024c) are inspired by analysis of Strauss and Yorke (1967a, b) and by Gripenberg et al. (1990, Lemma 15.9.2).

We now discuss the connection between (A) and (B) in Theorem 5. The fact that (B) implies (A) is due to Gripenberg et al. (1990), while Strauss and Yorke (1967a, b) established the necessity and sufficiency of (B), albeit with extra uniform (in θ) convergence required. As pointed out in Appleby and Lawless (2024a), it is easy to see (A) implies (B) by integrating across (14) and re–arranging:

$$\int_{t-\theta}^{t} f(s)\,ds = y(t) - y(t-\theta) + \int_{t-\theta}^{t} y(s)\,ds, \tag{17}$$

since the righthand side tends to zero as $t \to \infty$ if $y(t)$ does.

We state results characterising the rate of growth or decay of solutions of (14). We allow either for an increasing upper bound, or one which grows or decays to zero subexponentially (exponential or faster–than–exponential decay is not interesting, because solutions of (1) decay at a power law rate and swamp exponentially small terms). Recall that a function is $\gamma \in C((0,\infty);(0,\infty))$ is subexponential if

$$\lim_{t\to\infty} \frac{\gamma(t-\theta)}{\gamma(t)} = 1, \quad \theta > 0. \tag{18}$$

This is part of a definition of subexponential functions (see e.g., Bingham et al. 1991). For a comprehensive discussion of properties of this class of functions in the context of differential equations of the type studied in this paper, see Appleby and Lawless (2024c). We list some relevant properties to our investigation now. The convergence in (18) is in fact uniform on θ–compacts, so if $\Gamma(t) = \int_t^{t+1} \gamma(s)\,ds$, then $\Gamma(t) \sim \gamma(t)$ as $t \to \infty$ and moreover $\Gamma'(t)/\Gamma(t) \to 0$ as $t \to \infty$. By asymptotic integration of $\Gamma'(t)/\Gamma(t) \to 0$ as $t \to \infty$ we get $\log\Gamma(t)/t \to 0$ as $t \to \infty$, whence $\log\gamma(t)/t \to 0$ as $t \to \infty$; thus we get $\lim_{t\to\infty} e^{\epsilon t}\gamma(t) = +\infty$ and $\lim_{t\to\infty} e^{-\epsilon t}\gamma(t) = 0$ for all $\epsilon > 0$. These limits justify the terminology subexponential. Lastly, by using l'Hôpital's rule, the above remarks show, for any $\alpha > 0$, that

$$\int_0^t e^{-\alpha(t-s)}\gamma(s)\,ds \sim \frac{1}{\alpha}\Gamma(t) \sim \frac{1}{\alpha}\gamma(t) \quad t \to \infty. \tag{19}$$

The next result from Appleby and Lawless (2024c) characterises of growth or decay rates of the solution (14), if the upper bound is monotone or subexponential. The crucial object to study is the function $f_\theta(t) := \int_{(t-\theta)^+}^{t} f(s)\,ds$, $t \geq 0, \theta \in [0, 1]$; we compare it to a weight function γ via hypotheses such as (4),

$$f_\theta(t) = o(\gamma(t)) \text{ as } t \to \infty; \tag{20}$$

$$\text{There exists } \theta' > 0 \text{ such that } \limsup_{t\to\infty} |f_\theta(t)|/\gamma(t) > 0. \tag{21}$$

Theorem 6 *Let $f \in C([0,\infty);\mathbb{R})$, y be the unique continuous solution to* (14). *Let either γ be (a) positive, non–decreasing and in $C(0,\infty)$ or (b) subexponential. Then*

(i) (4) $\Longleftrightarrow y(t) = O(\gamma(t))$, *as* $t \to \infty$;
(ii) (20) $\Longleftrightarrow y(t) = o(\gamma(t))$, *as* $t \to \infty$;
(iii) (4) *and* (21) $\Longleftrightarrow \limsup_{t\to\infty} |y(t)|/\gamma(t) \in (0,\infty)$.

The hypothesis (A) in the above Theorem requires uniform control of f_θ for all $\theta \in [0,1]$. The choice of $[0,1]$ is arbitrary; control on an interval of the form $[0,\delta]$ for any choice of $\delta > 0$ is equivalent. This holds for all hypotheses of this type, whether the weight function γ is subexponential, regularly varying or monotone.

We see that the last part Theorem 6 can be used to characterise the size of solutions of (14) *exactly* to leading order. Roughly, $y(t)$ is $O(\gamma(t))$ but not $o(\gamma(t))$ as $t \to \infty$ if and only if $f_\theta(t)$ is $O(\gamma(t))$ for all θ, but not $o(\gamma(t))$ for some θ, as $t \to \infty$.

4 Asymptotics of Deterministic Pantograph Equation

It is now easy to characterise boundedness of solutions of the pantograph equation.

Theorem 7 *Suppose that $q \in (0,1)$ and* (2) *holds. Let $f \in C([0,\infty);\mathbb{R})$ and x be the unique continuous solution to* (7). *Then the following are equivalent*

(A) *y is bounded;*
(B) *There exists $F \ge 0$ such that $|\int_t^{t+\theta} f(s)\,ds| \le F$ for all $\theta \in [0,1]$, $t \ge 0$;*
(C) *x is bounded.*

To prove this, apply part (A) of Theorem 4 to $z = x - y$. Since y is bounded, φ given by (15) is bounded. Thus z is bounded and hence $x = z + y$ is bounded, so (A) implies (C). (C) implies (A) by virtue of (16). Applying part (i) of Theorem 6 with $\gamma(t) = 1$ shows that (A) and (B) are equivalent, completing the proof.

We note of course that taking the results of Theorems 5 and 7 together characterise when x is bounded but non–convergent. This will happen if and only if condition (B) of Theorem 7 holds, but condition (B) of Theorem 5 does not hold.

A simple situation where this happens is if $f(t) \to f^* \neq 0$ as $t \to \infty$, or if $f(t) = \sin(t)$ for $t \ge 0$. A more complicated situation, where f is unbounded, is furnished by a simple modification of the second example after Theorem 5. To give boundedness (but non–convergence) of solutions of the perturbed pantograph equation, we now choose $w_n = 1/h_n$, while still taking $h_n > 1$ and $h_n \uparrow \infty$ as $n \to \infty$ (once again, therefore, the rate of growth of the running maximum of $|f|$ can be arbitrarily fast in time). In this case, for all $\theta \in [0,1]$ and all $t \ge 0$ we have

$$\int_t^{t+\theta} f(s)\,ds \le \int_t^{t+1} f(s)\,ds \le \int_n^{n+2} f(s)\,ds = 1,$$

where n is the largest integer less than or equal to t. Thus condition (B) of Theorem 7 holds. On the other hand

$$\int_n^{n+1} f(s)\,ds = \frac{1}{2},$$

for every integer n, so condition (B) of Theorem 5 does not hold.

An interesting class of bounded solutions are those which tend to non–trivial limits. The situation when this happens can also be characterised. Once again, we believe this is the first such characterisation of convergent solutions for nonautonomous functional differential equations. We leave to the interested reader the proof of the following theorem, whose proof is similar to those of Theorems 5 and 7. Consult the proof of Gripenberg et al. (1990, Lemma 15.9.2) to prove the equivalence of (A) and (B); the equivalence of (A) and (C) can be proven with sole recourse to arguments given here.

Theorem 8 *Suppose that* $q \in (0, 1)$ *and* (2) *holds. Let* $f \in C([0, \infty); \mathbb{R})$ *and* x *be the unique continuous solution to* (7). *Then the following are equivalent*

(A) $y(t)$ *tends to a non–trivial finite limit as* $t \to \infty$;
(B) There exists $F \neq 0$ *such that for every* $\theta \in [0, 1]$ *that* $\frac{1}{\theta}\int_t^{t+\theta} f(s)\,ds \to F$ *as* $t \to \infty$ *for all* $\theta \in [0, 1]$;
(C) $x(t)$ *tends to a non–trivial finite limit as* $t \to \infty$.

It can be easily shown that the limits at infinity of x and y are related to F in condition (B) according to

$$\lim_{t\to\infty} x(t) = \frac{F}{1-(b+a)}, \quad \lim_{t\to\infty} y(t) = F.$$

We now use Theorem 2 to give rather sharp conditions on the asymptotic behaviour of solutions, without imposing pointwise conditions on the forcing function f. We start by giving a theorem for large perturbations.

Theorem 9 *Suppose that* $q \in (0, 1)$ *and* (2) *holds. Let* $f \in C([0, \infty); \mathbb{R})$ *and* x *be the unique continuous solution to* (7). *Suppose* γ *obeys either (i)* $\gamma \in RV_\infty(\eta)$ *for some* $\eta > \kappa$ *or (ii)* γ *is positive, increasing and in* C^1, *with* $\gamma(t) \to \infty$ *as* $t \to \infty$.

(I) (4) $\Longleftrightarrow y(t) = O(\gamma(t))$, *as* $t \to \infty \Longleftrightarrow x(t) = O(\gamma(t))$, *as* $t \to \infty$.
(II) (20) $\Longleftrightarrow y(t) = o(\gamma(t))$, *as* $t \to \infty \Longleftrightarrow x(t) = o(\gamma(t))$, *as* $t \to \infty$.
(III) (4), (21) $\Leftrightarrow \limsup_{t\to\infty} |y(t)|/\gamma(t) \in (0, \infty) \Leftrightarrow \limsup_{t\to\infty} |x(t)|/\gamma(t) \in (0, \infty)$.

Proof Note part (III) follows from parts (I) and (II). For part (I), note that if γ is regularly varying, it is also subexponential, so (A) and (B) are equivalent by Theorem 6. Thus is suffices to prove the equivalence of (B) and (C).

Next, let $z = x - y$ and $\varphi(t) = (1 + b)y(t) + ay(qt)$. Then z obeys (10). If (B) holds, $|y(t)| \le C_0\gamma(t)$. Then, as γ is regularly varying, $|\varphi(t)| \le C\gamma(t)$ for some

$C > 0$. Since $\eta > \kappa$, part (C) of Theorem 2 applies, so $z(t) = O(\gamma(t))$ as $t \to \infty$. But since $y(t) = O(\gamma(t))$ and $x = y + z$, $x(t) = O(\gamma(t))$ as $t \to \infty$, which is (C).

To show (C) implies (B), note that (C) yields $|x(t)| \le K\gamma(t)$ for $t \ge 0$. This also gives $|x(qt)| \le K\gamma(qt) \le L\gamma(t)$, using the regular variation of γ, for constants K and L. Applying the triangle inequality to (16), and using these estimates, we get

$$|y(t)| \le |\xi|e^{-t} + K\gamma(t) + \int_0^t e^{-(t-s)}\left(|1+b|K\gamma(s) + |a|L\gamma(s)\right) ds. \tag{22}$$

Now, note that there is a Γ such that $\gamma \sim \Gamma$ and $t\Gamma'(t) \sim \eta\Gamma(t)$ as $t \to \infty$. Thus, $\Gamma'(t) = o(\Gamma(t))$ as $t \to \infty$ and $e^t\Gamma(t) \to \infty$ as $t \to \infty$. Thus the integral in (22) is majorised by some positive constant times $e^{-t}\int_0^t e^s\Gamma(s)\,ds$. By L'Hôpital's rule,

$$\lim_{t\to\infty} \frac{\int_0^t e^s\gamma(s)\,ds}{e^t\Gamma(t)} = \lim_{t\to\infty} \frac{e^t\Gamma(t)}{e^t\Gamma(t) + e^t\Gamma'(t)} = \lim_{t\to\infty} \frac{1}{1+\Gamma'(t)/\Gamma(t)} = 1.$$

Since $\gamma \sim \Gamma$, the integral in (22) is $O(\gamma(t))$, and thus $y(t) = O(\gamma(t))$, which is (B).

For part (II) in the regularly varying case, note that if γ is regularly varying, it is also subexponential, so (A) and (B) are equivalent by Theorem 6. Thus it suffices to prove the equivalence of (B) and (C). Once again, let $z = x - y$ and $\varphi(t) = (1+b)y(t) + ay(qt)$. Then z obeys (10) and $\varphi(t) = o(\gamma(t))$ as $t \to \infty$, by hypothesis. Revisiting the proof of part (C) of Theorem 2, it can be shown in a similar way that $z(t) = o(\gamma(t))$ as $t \to \infty$. Hence $x(t) = o(\gamma(t))$ as $t \to \infty$, so (B) implies (C). To show that (C) implies (B), note by hypothesis that $|x(t)| < \epsilon\gamma(t)$ as $t \to \infty$ for all $t \ge T$. Using the regular variation, this also gives $|x(qt)| \le C\epsilon\gamma(t)$ for all $t \ge T/q = T_1$ and some t and ϵ independent $C > 0$. Now take $t \ge T_1 > T$, and applying the triangle inequality to (16), we have

$$\begin{aligned} |y(t)| \le |\xi|e^{-t} + \epsilon\gamma(t) + e^{-t}\left|\int_0^{T_1} e^s((1+b)x(s) + ax(qs))\,ds\right| \\ + \epsilon\left(|1+b| + |a|C\right)\int_{T_1}^t e^{-(t-s)}\gamma(s)\,ds. \end{aligned}$$

Dividing by $\gamma(t)$ and taking the limit as $t \to \infty$ on the righthand side, we get limits ϵ, 0 and $\epsilon\,(|1+b| + |a|C)$ respectively for the three terms, using the fact that γ is subexponential in the last case. Thus $\limsup_{t\to\infty} |y(t)|/\gamma(t) \le \epsilon(1 + |1+b| + |a|C)$, so since ϵ is arbitrary, we have $y(t) = o(\gamma(t))$ as $t \to \infty$, as needed.

For part (I) in the monotone case, (A) and (B) are equivalent by Theorem 6, so if (B) and (C) are equivalent, we are done. Let $z = x - y$ and $\varphi(t) = (1+b)y(t) + ay(qt)$. Then z obeys (10). Suppose (B) holds, so $|y(t)| \le C\gamma(t)$. Then, as γ is increasing, $|\varphi(t)| \le C(|1+b| + |a|)\gamma(t)$. Hence by Theorem 4, $z(t) = O(\gamma(t))$. But since $y(t) = O(\gamma(t))$ and $x = y + z$, we have $x(t) = O(\gamma(t))$ as $t \to \infty$, or (B). To show (C) implies (B), note from (C) that $|x(t)| \le K\gamma(t)$ for $t \ge 0$. Applying the triangle inequality to (16), using this estimate, and using the monotonicity of γ, we get

$$\begin{aligned}|y(t)| &\leq |\xi|e^{-t} + K\gamma(t) + \int_0^t e^{-(t-s)} \left(|1+b|K\gamma(s) + |a|K\gamma(qs)\right) ds \\ &\leq |\xi|e^{-t} + K\gamma(t) + \int_0^t e^{-(t-s)} \left(|1+b|K + |a|K\right) ds \cdot \gamma(t) \\ &\leq (|\xi|/\gamma(0) + K + |1+b|K + |a|K)\gamma(t),\end{aligned}$$

which is (B). The proof of (II) in the monotone case uses ideas from parts (I) in the monotone and (II) in the regularly varying case, which is left to the reader. □

For smaller perturbations, it is hard to prove as clean a result. If γ is regularly varying, and $x(t) = O(\gamma(t))$, using (16) yields $y(t) = O(\gamma(t))$ as $t \to \infty$. In particular, take $\gamma(t) = t^\kappa$. However, this over–estimates the size of the smallest perturbation needed to preserve the rate of decay t^κ of (1), since we know that if $y(t) = O(\gamma(t))$ where $\psi(t) := \gamma(t)/t^{\kappa+1}$ is regularly varying with index -1, and is integrable, then $x(t) = O(t^\kappa)$ as $t \to \infty$. But the integrability of ψ means that $t\psi(t) \to 0$ as $t \to \infty$, and so the largest perturbations which preserve the rate of decay are $o(t^\kappa)$ (without further restrictions being placed on y).

Theorem 10 *Suppose that $q \in (0, 1)$ and (2) holds. Let $f \in C([0, \infty); \mathbb{R})$ and x be the unique continuous solution to (7). Let $\gamma \in RV_\infty(\eta)$.*

(i) Let $y(t) = O(\gamma(t))$ as $t \to \infty$.

(A) If $\eta < \kappa$, then $x(t) = O(t^\kappa)$ as $t \to \infty$.

(B) If $\eta = \kappa$ and $t \mapsto \gamma(t)/t^{\kappa+1}$ is in L^1, then $x(t) = O(t^\kappa)$ as $t \to \infty$. Moreover, if $\limsup_{t\to\infty} |y(t)|/\gamma(t) \in (0, \infty)$, then $\limsup_{t\to\infty} |x(t)|/\gamma(t) > 0$.

(C) If $\eta = \kappa$ and $t \mapsto \gamma(t)/t^{\kappa+1}$ is not in L^1, then

$$x(t) = O\left(t^\kappa \int_1^t \frac{\gamma(s)}{s^{\kappa+1}} ds\right), \quad t \to \infty.$$

Moreover $\limsup_{t\to\infty} |y(t)|/\gamma(t) > 0$ implies $\limsup_{t\to\infty} |x(t)|/\gamma(t) > 0$.

(ii) If $x(t) = O(\gamma(t))$ as $t \to \infty$, then $y(t) = O(\gamma(t))$ as $t \to \infty$. Likewise, if $x(t) = o(\gamma(t))$ as $t \to \infty$, then $y(t) = o(\gamma(t))$ as $t \to \infty$.

Notice that the hypotheses on y required to classify the asymptotic behaviour of x are all equivalent to hypotheses on f, as laid out in Theorem 9. Proofs of upper bounds in part (i) follow by considering $z = x - y$, noting that $y = O(\gamma)$ implies $\varphi = O(\gamma)$ and then applying Theorem 2 to z. The proofs of lower bounds are by contradiction. Note first that (ii) can be shown by using (16). Then, contradiction of the desired conclusion leads to supposing that $x(t) = o(\gamma(t))$. But then $y(t) = o(\gamma(t))$ by (16), contradicting the hypothesis that $\limsup_{t\to\infty} |y(t)|/\gamma(t) > 0$.

We now make some remarks. First, observe that, taken together, Theorems 10 and 9 cover a very wide class of perturbation sizes, and are in fact very comprehensive for practical purposes.

Second, note that part (A) and (B) of the theorem are in the form of a Hartman–Wintner type–result (see e.g., Hartman and Wintner 1955 for the first results by Hartman (2002, Cor X.16.4) for a typical result on ordinary equations and for linear functional equations, see Pituk (1999)). A typical feature of Hartman–Wintner results is that if perturbation terms obeys some integral smallness condition, then the solution of the perturbed equation x can be related to the solution of the unperturbed problem in the form $x = O(u)$, or perhaps x/u tends to a finite limit; we see therefore that parts (A) and (B) of the theorem have this character. Clearly, Hartman–Wintner results are a strong type of "asymptotic preservation" result.

A weaker, but still very useful asymptotic preservation result, is a Perron-type result (for a typical result for linear, or linearisable, autonomous functional differential equations, see e.g., Pituk (2006)). In that case, we show that some nonlinear function of the solution behaves like a known function of t. This allows us ignore dependence on initial data in bounds or limits (which can be hard to quantify explicitly anyway), and enables us to admit more irregularity (or to get away with weaker estimation) in the perturbed terms. For linear equations (or equations that can be linearised) Perron–type results often concentrate on (top) Liapunov exponents i.e. $\limsup_{t\to\infty} \log|x(t)|/t$ or $\lim_{t\to\infty} \log|x(t)|/t$, if the limit exists. We see that the existence of these limits is weaker than knowing that $x(t) \sim ce^{\lambda t}$ as $t \to \infty$ for some $c \neq 0$, and this is the price that must typically be paid in Perron–type results. In our case, because the solution has power–law asymptotic behaviour in the unperturbed case, we consider $\limsup_{t\to\infty} \log|x(t)|/\log t$, noting that for almost all solutions of the unperturbed equation we have

$$\limsup_{t\to\infty} \frac{\log|u(t)|}{\log t} = \kappa.$$

We also notice that this is consistent with classic ideas of polynomial asymptotic stability in stochastic equations, due to Mao (1992a, b) and generalisations to other (in general, non–exponential) decay rates; see e.g., Wu and Hu (2012). In addition, we notice that considering solutions on this logarithmic scale is implicit in the bounds obtained for general delay differential equations in Čermák (2002).

By focussing on the size of $\log|x(t)|$ in this way, we may characterise solution sizes very cleanly in terms of y, for both large and small perturbations, in the following Perron–type theorem.

Theorem 11 *Suppose that* $q \in (0, 1)$ *and* (2) *holds. Let* $f \in C([0,\infty);\mathbb{R})$ *and* x *be the unique continuous solution to* (7). *Let* $|\theta(t)| \nearrow \infty$ *as* $t \to \infty$, *and suppose* $\theta(t)/\log t \to \eta \in [\kappa, \infty]$ *as* $t \to \infty$.

(A) If $\eta = \kappa$, then

$$\limsup_{t\to\infty} \frac{\log|y(t)|}{\theta(t)} \leq 1 \iff \limsup_{t\to\infty} \frac{\log|x(t)|}{\theta(t)} \leq 1.$$

(B) If $\eta \in (\kappa, \infty]$, then

$$\limsup_{t\to\infty} \frac{\log|y(t)|}{\theta(t)} = 1 \iff \limsup_{t\to\infty} \frac{\log|x(t)|}{\theta(t)} = 1.$$

Part (A) is equivalent to the more easily–understood statement

$$\limsup_{t\to\infty} \frac{\log|y(t)|}{\log t} \leq \kappa \iff \limsup_{t\to\infty} \frac{\log|x(t)|}{\log t} \leq \kappa,$$

and part (B), for finite $\eta > \kappa$, is equivalent to

$$\limsup_{t\to\infty} \frac{\log|y(t)|}{\log t} = \eta \iff \limsup_{t\to\infty} \frac{\log|x(t)|}{\log t} = \eta,$$

but we have stated Theorem 11 in the form above to show unity between the small and large perturbation cases, irrespective of the solution or perturbation size.

To prove part (B) of Theorem 11, we break it into the cases when $\eta \in (\kappa, \infty)$ and $\eta = \infty$. We prove the former case, leaving the similar proof in the latter case to the interested reader. The reader will immediately note that the following proof requires mainly a re–interpretation or re–working of parts of earlier results, but besides this, no important new analysis is needed.

If $\eta > \kappa$, the first statement in (B) means that for every $\epsilon > 0$ there is a $C_\epsilon > 0$ such that $|y(t)| \leq C_\epsilon t^{\eta+\epsilon}$ for all $t \geq 1$. Because $\eta > \kappa$, we can apply Theorem 9 to give $|x(t)| \leq D_\epsilon t^{\eta+\epsilon}$ for all $t \geq 1$ and some $D_\epsilon > 0$. As a result, we get that $\limsup_{t\to\infty} \log|x(t)|/\log t \leq \eta + \epsilon$. Letting $\epsilon \to 0$, we get that $\limsup_{t\to\infty} \log|x(t)|/\theta(t) \leq 1$. Suppose now, by way of contradiction, that the limsup is strictly less than unity; call it l. Then, for any $\epsilon > 0$ so small that $(l+\epsilon)(\eta+\epsilon) < \eta$, there exists a $T(\epsilon) > 0$ such that we have $\log|x(t)| < (l+\epsilon)\theta(t) < (l+\epsilon)(\eta+\epsilon)\log t$ for $t \geq T(\epsilon)$. Therefore, by Theorem 9, we have that there exists a $C_\epsilon > 0$ such that $|y(t)| \leq C_\epsilon t^{(l+\epsilon)(\eta+\epsilon)}$ for all $t \geq 1$. Now, take logarithms, divide by $\log t$ and let $t \to \infty$; and noting that the inequality holds for all $\epsilon > 0$ sufficiently small, we may further let $\epsilon \to 0^+$, giving

$$\limsup_{t\to\infty} \frac{\log|y(t)|}{\log t} \leq l\eta < \eta,$$

while by hypothesis, this limsup is η, yielding the desired contradiction. Hence the forward implication holds.

For the backward implication, the hypothesis implies that for every $\epsilon > 0$ there is a $C_\epsilon > 0$ such that $|x(t)| \leq C_\epsilon t^{\eta+\epsilon}$ for all $t \geq 1$. By (16) this implies that there exists $D_\epsilon > 0$ such that $|y(t)| \leq D_\epsilon t^{\eta+\epsilon}$ for $t \geq 1$, and so, by taking logs, the limit as $t \to \infty$, and finally the limit as $\epsilon \to 0$ as before, we have $\limsup_{t\to\infty} \log |y(t)|/\theta(t) \leq 1$. As before, assume by way of contradiction that the limsup l is strictly less than unity. Then for all ϵ sufficiently small we have $l + \epsilon < 1$, and $(l+\epsilon)(\eta+\epsilon) < \eta$, and consequently for such an $\epsilon > 0$ there is a $T(\epsilon) > 0$ such that $\log |y(t)| < (l + \epsilon)\theta(t) < (l+\epsilon)(\eta+\epsilon)\log t$ for all $t \geq T(\epsilon)$.

In the case that $l\eta < \kappa$, we may choose $\epsilon > 0$ so that $(l+\epsilon)(\eta+\epsilon) \leq \kappa$; then by Theorem 10, we have that $x(t) = O(t^\kappa \log t)$, which means that $\limsup_{t\to\infty} \log x(t)/\log t \leq \kappa$. But $\limsup_{t\to\infty} \log |x(t)|/\log t = \eta$, so $\eta \leq \kappa$, which is false, and generates a contradiction in that case.

In the other case, where $l\eta \geq \kappa$, for every $\epsilon > 0$ we have $(l+\epsilon)(\eta+\epsilon) > \kappa$; in addition choose ϵ so small that $l + \epsilon < 1$, and $(l+\epsilon)(\eta+\epsilon) < \eta$. Then by Theorem 9 we have that $|x(t)| \leq D_\epsilon t^{(l+\epsilon)(\eta+\epsilon)}$, which reduces to

$$\limsup_{t\to\infty} \frac{\log |x(t)|}{\log t} \leq (l+\epsilon)(\eta+\epsilon).$$

By hypothesis, this limsup is η, so $\eta \leq (l+\epsilon)(\eta+\epsilon)$. But $(l+\epsilon)(\eta+\epsilon) < \eta$, which generates the desired contradiction. We have thus shown (B) in the case where $l\eta \geq \kappa$.

The proof of part (A), where $\eta = \kappa$, is very similar to the argument given above, and again we leave it to the interested reader to complete.

The clean picture generated by the combination of the Hartman–Wintner type Theorem 10, the large perturbation characterisation Theorem 9 and the Perron–type characterisation Theorem 11 is spoiled by the inequality in part (A) of Theorem 11, and the converse part (ii) in Theorem 9. This converse does not give a result like

$$x(t) = O(t^\kappa),\ t \to \infty \Longrightarrow y(t) = O(\gamma(t)),\ t \to \infty \text{ where } t \mapsto \gamma(t)/t^{\kappa+1} \in L^1(0,\infty),$$

which, together with Theorem 10, would essentially characterise the maximal allowable size of perturbations; instead, the best we can do in the present work is to show that but rather $x(t) = O(t^\kappa)$ as $t \to \infty$ implies $y(t) = O(t^\kappa)$ as $t \to \infty$.

Some scrutiny of part (ii) of Theorem 10 may suggest a reason why a clean converse may be difficult to obtain. It suggests that if a solution of the perturbed equation decays faster than the unperturbed equation (1), this can only be achieved with a rapidly vanishing perturbation. It is not difficult to manufacture examples of this phenomenon. Let μ be a positive, continuous and integrable function. Rescale μ so that $\int_0^\infty \mu(s)\,ds = 1$. Let $x(0) = \xi \neq 0$ and suppose f is given by

$$f(t) = \xi\left(-\mu(t) - b\int_t^\infty \mu(s)\,ds - a\int_{qt}^\infty \mu(s)\,ds\right), \quad t \geq 0.$$

Then $x(t) = \xi \int_t^\infty \mu(s)\,ds$ for $t \geq 0$ is the unique continuous solution of (7). If we suppose that μ tends to zero superexponentially, in the sense that $\lim_{t\to\infty} \log \mu(t)/t = -\infty$, then x tends to zero superexponentially, and so does f; moreover, the more rapid the decay in f, the more rapid the decay in x. Thus, we may have the situation that the equation may be is perturbed by two "small" perturbations, but that the smaller perturbation gives rise to a more slowly decaying solution. For instance, if $a > 0$, $x(0) > 0$ and $f(t) \equiv 0$, we have that there exist $0 < c_1 \leq c_2 < +\infty$ such that $c_1 t^\kappa \leq x(t) \leq c_2 t^\kappa$ for all $t \geq 1$, while the solution in the above example, where there is a non–trivial perturbation, decays superexponentially fast when $\log \mu(t)/t \to -\infty$ as $t \to \infty$.

The reader may also wonder whether the upper bound result in parts (i)(A) and (i)(B) of Theorem 10 can be improved in at least some situations to the stronger asymptotic estimate $x(t) \sim ct^\kappa$ for some $c \neq 0$. One situation where this seems especially plausible is that in which f is positive, and $a > 0$, since these conditions promote the positivity of the solution, and would tend to mitigate against the oscillation in the solution that is exhibited when $a < 0$ and $f \equiv 0$. However, even in these situations where positivity of the solution is promoted, it is still not always possible to rule out the situation that $x(t)/t^\kappa$ is a bounded function which does not tend to a limit. Therefore, Theorem 10 is also sharp in terms of its conclusions; no better order estimate is available, granted the hypotheses imposed therein.

To see that we cannot always expect solutions to be asymptotic to t^κ as $t \to \infty$, consider the function x defined by

$$x(t) = t^\kappa \left\{ C + \sin\left(\frac{2\pi}{\log(1/q)} \log t \right) \right\} + Dt^\kappa \int_q^t \psi(s)\,ds, \quad t \geq q,$$

where $\psi \in RV_\infty(-1)$, $\psi \in L^1(0,\infty)$, and $C > 1$, $D > 0$. Let $a > 0$, $b < 0$ and $-b > a$. Let $\pi(t) := C + \sin(2\pi \log t/\log(1/q))$ for $t \geq q$. Define $f(t) = x'(t) - bx(t) - ax(qt)$ for all $t \geq 1$. Then x solves (7) on $t \geq 1$ and $x(t) > 0$ for all $t \geq q$. Moreover $x(t)/t^\kappa$ is bounded, but does not tend to a limit as $t \to \infty$. We now show that f is asymptotically positive and satisfies the conditions of part (i)(A) of Theorem 10.

Note that $\pi(qt) = \pi(t)$ for all $t \geq 1$, so using $b + aq^\kappa = 0$ and the uniform convergence theorem, we have that

$$\alpha(t) := -(bx(t) + ax(qt)) = |b|Dt^\kappa \int_{qt}^t \psi(s)\,ds \sim |b|D\psi(t)t^{\kappa+1}\log(1/q), \quad t \to \infty.$$

The function on the righthand side is positive, in $\mathrm{RV}_\infty(\kappa)$, but is $o(t^\kappa)$ as $t \to \infty$, because the integrability and asymptotic monotonicity of ψ (the latter property follows from the regular variation of ψ) imply $t\psi(t) \to 0$ as $t \to \infty$.

On the other hand, a direct calculation yields

$$x'(t) = t^{\kappa-1}\left(\kappa\pi(t) + \frac{2\pi}{\log(1/q)}\cos\left(\frac{2\pi}{\log(1/q)}\log t\right) + Dt\psi(t) + \kappa D\int_q^t \psi(s)\,ds\right).$$

Note that the term in the brackets is uniformly bounded, since π is bounded, ψ is integrable, and $t\psi(t) \to 0$ as $t \to \infty$. Hence $x'(t) = O(t^{\kappa-1})$ as $t \to \infty$. Therefore, as α is asymptotic to a $RV_\infty(\kappa)$ function, $x'(t) = o(\alpha(t))$ as $t \to \infty$. Thus $f(t) = x'(t) + \alpha(t) \sim |b|D\log(1/q)\psi(t)t^{\kappa+1}$ as $t \to \infty$. From this it is easy to show that $y(t) \sim f(t)$ as $t \to \infty$. Therefore, we may take $\gamma(t) := |b|D\log(1/q)\psi(t)t^{\kappa+1}$ in Theorem 10 because it is in $\mathrm{RV}_\infty(\kappa)$. Moreover, since ψ is integrable,

$$\int_1^T \frac{\gamma(s)}{s^{\kappa+1}}\,ds = |b|D\log(1/q)\int_1^T \psi(s)\,ds \to L, \quad T \to \infty,$$

so although y fulfills the properties of part (i)(B) of Theorem 10, with $a > 0$ and f positive for all t sufficiently large, $t \to x(t)/t^\kappa$ does not tend to a constant as $t \to \infty$. This demonstrates, as claimed above, that the "big O" result in Theorem 10 is the best possible in general, given the hypotheses imposed here. This example also hints at the possible asymptotic form of the bounded component of $x(t)/t^\kappa$, under appropriate smallness conditions on y, and we propose to investigate this further in later works.

A characterisation, purely in terms of $f_\theta(t)$, of the situation wherein $\log|y(t)|$ has certain upper rates of growth or decay, has not been presented here. However, it seems that such a result should be available, and would be especially useful when considering Perron–type results for differential equations with general unbounded delay.

Next, we compare the results of Theorems 9 and 10 with perturbation theorems of Cermak (see e.g., Theorem 2.3 and Example 4.1 in Čermák 2002), which, in our view provide the sharpest and most comprehensive estimates for the asymptotic behaviour of perturbed functional differential equations with unbounded delay. He supposes that $f(t) = O(t^\beta)$ and $f'(t) = O(t^{\beta-1})$ as $t \to \infty$. Applying Cermak's results to the pantograph equation, we get

$$\begin{aligned}
x(t) &= O(t^\kappa), \quad t \to \infty, \quad \text{if } \beta < \kappa,\\
x(t) &= O(t^\kappa \log t), \quad t \to \infty, \quad \text{if } \beta = \kappa,\\
x(t) &= O(t^\beta), \quad t \to \infty, \quad \text{if } \beta > \kappa.
\end{aligned}$$

First, we observe that our results need neither need pointwise estimates on f, nor on f', but merely on y, where $y' = -y + f$, and the smoothness of f is not needed in our results. Second, it can readily be the case that f has very bad pointwise behaviour, yet y has good behaviour. For example, suppose that

$$f(t) = t^\beta + e^t\sin(e^{2t}), \quad t \geq 0,$$

Clearly, $f(t) = O(e^t)$ and $f'(t) = O(e^{3t})$ as $t \to \infty$, yet $y(t) = O(t^\beta)$ as $t \to \infty$. Thus, applying Theorem 10, we are able to conclude that

$$x(t) = O(t^{\max(\kappa,\beta)}), \quad t \to \infty, \text{for } \beta \neq \kappa$$

and $x(t) = O(t^\kappa \log t)$ as $t \to \infty$ if $\beta = \kappa$, while Cermak's theorem cannot give any conclusions.

Furthermore, the existing results require an explicit parametric comparison between the solution of the unperturbed equation and perturbation. For example, if $f(t) = O(t^\beta \log t)$ as $t \to \infty$, we cannot apply Theorem 2.3 in Čermák (2002) directly; instead, we would note that $f(t) = O(t^{\beta+\epsilon})$ for every $\epsilon > 0$, and hope to obtain the estimate $f'(t) = O(t^{\beta-1+\epsilon})$ on the derivative. Thus, if $\beta < \kappa$, we get $x(t) = O(t^\kappa)$ and if $\beta > \kappa$ we get $x(t) = O(t^\beta)$, but no prediction can be made in the case $\beta = \kappa$. Using Theorem 10 however, we get

$$x(t) = O\left(t^\kappa \int_1^t s^{\beta-\kappa-1} \log s \, ds\right), \quad t \to \infty.$$

The estimates concur with (Čermák 2002, Theorem 2.3) in the cases when $\beta \neq \kappa$; but when $\beta = \kappa$, we get

$$x(t) = O(t^\kappa \log^2(t)), \quad t \to \infty,$$

which is not covered by Cermak's theorem.

These considerations seem to show that the estimation in the Fundamental Lemma 1 may be slightly sharper in characterising the transition from small to large perturbations than the proof of Čermák (2002, Theorem 2.3). It is an interesting open question to revisit Cermak's delicate and beautiful proof of Čermák (2002, Lemma 3.3) for equations with general unbounded delay. This result underlies (Čermák 2002, Theorem 2.3), and is a close analogue of Lemma 1. This is especially worthwhile, because we have shown by means of the general examples in Theorem 3 that the estimates in Theorem 10 are, in a certain sense, unimprovable.

Finally, we note that Theorem 9 above deals with large perturbations; suppose that $y(t) = O(e^{\beta t})$ as $t \to \infty$ for some $\beta > 0$. Then this perturbation cannot be put into the framework of Čermák (2002, Theorem 2.3). However, by Theorem 9, we have that $x(t) = O(e^{\beta t})$ as $t \to \infty$. Furthermore, if it is the case that $\limsup_{t\to\infty} |x(t)|/e^{\beta t} \in (0, \infty)$, this can happen if and only if $\limsup_{t\to\infty} |y(t)|/e^{\beta t} \in (0, \infty)$. However, stronger conclusions about f are not available; for instance consider

$$f(t) = e^{\beta t} + ce^{\theta t} \sin(e^{3\theta t})$$

where $\theta > \beta > 0$. If $c = 0$, then $f(t) = O(e^{\beta t})$ and $y(t) = O(e^{\beta t})$ as $t \to \infty$. If $c \neq 0$, then we still have $y(t) = O(e^{\beta t})$ as $t \to \infty$, but $f(t) = O(e^{\theta t})$. Thus, both f's generate the same asymptotic size of y, and hence of x, but knowing the size of

x does not allow conclusions to be drawn about the order of f, merely that of y, or equivalently, by Theorem 6, the order of f_θ.

In the case of small perturbations i.e. when $y(t) = O(\gamma(t))$ as $t \to \infty$ and $t \mapsto \gamma(t)/t^{\kappa+1}$ in integrable, we have that $x(t) = O(t^\kappa)$ as $t \to \infty$, or to put it another way, $t \mapsto x(t)/t^\kappa$ is a bounded function. In the unperturbed case, the asymptotic structure of this bounded component has been found; it is the sum of a log–periodic function and a function which tends to zero as $t \to \infty$ (see Kato and McLeod 1971). It is the subject of further work to determine whether these small perturbation conditions are sufficient to establish that the bounded component of the perturbed equation retains a similar (log–periodic) structure.

Some preliminary ideas as to how this might be achieved are presented in Sect. 5 and at the end of the paper in Sect. 7. Space constraints, and the necessity to develop a rate characterisation for related stochastic differential equations (see the next section) preclude more exact analysis and discussion in this work.

Thus, we see that a correct determination of the exact asymptotic order is likely a first step in understanding the detailed asymptotic structure of solutions. Moreover, it opens up the prospect that the solution *might retain the fine structure of its solutions despite bad pointwise behaviour in forcing functions, irregular stochastic terms, and relatively large forcing terms (as measured by y)*. In other words, the refined asymptotic bounds presented in this section can be seen as a critical first step in establishing that the solution structure is robust to relatively large and irregular perturbations.

5 Stochastically Forced Equation and Auxiliary SDEs

In this section, we consider the stochastically perturbed pantograph equations

$$dX(t) = (bX(t) + aX(qt) + f(t))\,dt + \sigma(t)\,dB(t), \quad t \geq 0; \quad X(0) = \xi, \tag{23}$$

where once again a and b are real numbers and $q \in (0, 1)$. For simplicity, we assume ξ is non–random, f is continuous as before, and

$$\sigma \in C([0, \infty); \mathbb{R}). \tag{24}$$

Here $B = \{B(t) : t \geq 0\}$ is a standard Brownian motion. In order that a solution of (23) is well–defined and has appropriate properties, we work on a complete probability space $(\Omega, \mathcal{F}, \mathbb{P})$, where Brownian motion B generates its own natural filtration $\mathcal{F}^B(t) = \sigma(\{B(s) : 0 \leq s \leq t\})$ with $\mathcal{F}^B(t) \subset \mathcal{F}$ for each $t \geq 0$. Under these assumptions, (23) is considered a shorthand for the integral equation

$$X(t) = \xi + \int_0^t \{bX(s) + aX(qs) + f(s)\}\,ds + \int_0^t \sigma(s)\,dB(s), \quad t \geq 0$$

which has a unique, continuous, and ($\mathcal{F}^B(t)$–) adapted solution on $[0, \infty)$ (see e.g., Mao 2008).

In just the way the auxiliary ordinary differential equation (14) allows sharp results to be proven for the deterministically forced equation (7), solutions of related auxiliary stochastic (ordinary) differential equations enable us to characterise the asymptotic behaviour of solution of (23). Let Y_0 and Y be solutions of the SDEs

$$dY_0(t) = -Y_0(t)\,dt + \sigma(t)\,dB(t), \quad t \geq 0; \quad Y_0(0) = 0, \tag{25}$$

$$dY(t) = (-Y(t) + f(t))\,dt + \sigma(t)\,dB(t), \quad t \geq 0; \quad Y(0) = 0. \tag{26}$$

Both equations possess unique, continuous, adapted solutions. Recalling that y is the solution of (14), the representations of the solutions of (25) and (26) are

$$Y_0(t) = e^{-t} \int_0^t e^s \sigma(s)\,dB(s), \quad t \geq 0, \tag{27}$$

$$Y(t) = e^{-t} \int_0^t e^s f(s)\,ds + e^{-t} \int_0^t e^s \sigma(s)\,dB(s) = y(t) + Y_0(t), \quad t \geq 0. \tag{28}$$

We show momentarily that the pathwise asymptotic behaviour of (25) and (26) can be characterised very precisely. For now, we show how Y can be connected with X. Let $Z = X - Y$. Set $\phi_1(t) = ay(qt) + (1+b)y(t)$ for $t \geq 0$ and

$$\phi(t) = aY(qt) + (1+b)Y(t), \quad t \geq 0 \tag{29}$$

so $\phi(t) = \phi_0(t) + \phi_1(t)$ where $\phi_0(t) = aY_0(qt) + (1+b)Y_0(t)$ and

$$Z'(t) = aX(qt) + bX(t) + Y(t) = aZ(qt) + bZ(t) + \phi(t), \quad t \geq 0. \tag{30}$$

Thus, we can apply the machinery of the previous section to Z, once the asymptotic behaviour of ϕ is known. This will be the case, since we will good information about the asymptotic behaviour of y and Y_0, on which ϕ entirely depends. Furthermore, since $X = Z + Y$, the asymptotic behaviour of X can be found.

In particular, suppose $b < 0$ and $|a| < |b|$ and that $Y(t) \to 0$ as $t \to \infty$ a.s. Then $\phi(t) \to 0$ as $t \to \infty$ a.s. Applying Theorem 1 part (a), we see from (30) that $Z(t) \to 0$ as $t \to \infty$ a.s. Thus, $X(t) = Z(t) + Y(t) \to 0$ as $t \to \infty$ a.s. In the case that Y is a.s. bounded, ϕ is bounded a.s., and this leads to Z being bounded a.s., by applying part (b) of Theorem 1 to (30). Hence $X = Z + Y$ is bounded a.s.

Converses result by writing Y in terms of X. Write $dX(t) = -X(t) + f(t) + \{(1+b)X(t) + aX(qt)\}\,dt + \sigma(t)\,dB(t)$. By stochastic integration by parts, we get

$$Y(t) = X(t) - \xi e^{-t} - \int_0^t e^{-(t-s)}\{(1+b)X(s) + aX(qs)\}\,ds, \quad t \geq 0. \tag{31}$$

From (31), it is clear that if $X(t) \to 0$ as $t \to \infty$, then $Y(t) \to 0$ as $t \to \infty$ a.s., and likewise, if X is bounded a.s., then Y is bounded a.s.

The above discussion shows the importance of possessing good asymptotic information concerning Y_0 and Y. Fortunately, some of this is already known for Y_0, and using known results, more can be established readily from scratch for Y. The following was established concerning Y_0 in Appleby et al. (2011), generalising results from Chan and Williams (1989).

Theorem 12 *Suppose σ obeys* (24). *Define $S(\epsilon)$ for $\epsilon > 0$ by* (3) *(if the sum diverges, we take $S(\epsilon) = +\infty$) and let Y_0 be the unique continuous adapted solution of* (25).

(a) $S(\epsilon) < +\infty$ *for all* $\epsilon > 0 \iff Y_0(t) \to 0$ *as* $t \to \infty$ *a.s.*
(b) There exists $\epsilon' > 0$ *such that* $S(\epsilon) < +\infty$ *for all* $\epsilon > \epsilon' \iff Y_0$ *is bounded a.s.*
(c) (There exists $\epsilon' > 0$ *such that* $S(\epsilon) < +\infty$ *for all* $\epsilon > \epsilon'$ *and* $S(\epsilon) = +\infty$ *for all* $\epsilon < \epsilon'$*)* $\iff 0 < \limsup_{t\to\infty} |Y_0(t)| < +\infty$ *a.s.*
(d) $S(\epsilon) = +\infty$ *for all* $\epsilon > 0 \iff \limsup_{t\to\infty} |Y_0(t)| = +\infty$ *a.s.*

It can be shown that S can only obey assumption (A) in part (a), (c) and (d) of the theorem, so solutions of (25) either tend to zero with probability one, are bounded but non–convergent with probability one, or are unbounded with probability one.

Notice the asymptotic behaviour of σ, measured through the function $\sigma_1^2(t)$ entirely governs the asymptotic behaviour of Y_0. This mirrors the manner in which the asymptotic decay or boundedness of $f_\theta(t)$ governs the convergence or boundedness of solutions of the deterministic equations (14) and (7).

It is also shown in Appleby et al. (2011) that if $\sigma_1^2(n)$ is asymptotic to a monotone (non–increasing) sequence, $Y_0(t) \to 0$ as $t \to \infty$ a.s. if and only if $\sigma_1^2(n)\log n \to 0$ as $n \to \infty$. In rough terms, therefore, $\sigma_1^2(t)$ must tend to zero faster than $1/\log t$ in order for $Y_0(t)$ to tend to zero as $t \to \infty$, a.s.

We now give a characterisation of the a.s. convergence and boundedness of Y. In broad terms, we have convergence to zero (respectively, boundedness) for the "doubly" perturbed equation if and only if both the "singly" perturbed equations tend to zero (respectively, are bounded).

Theorem 13 *Let f, σ obey* (6) *and* (24). *Let Y be the unique continuous adapted solution to* (26), *and suppose S is given by* (3).

(a) The following are equivalent:

- *(A) There exists* $\Delta > 0$ *such that* $f_\theta(t) \to 0$ *as* $t \to \infty$ *for all* $\theta \in [0, \Delta]$, $S(\epsilon) < +\infty$ *for all* $\epsilon > 0$*;*
- *(B)* $Y(t) \to 0$ *as* $t \to \infty$ *a.s.*

(b) The following are equivalent:

(A) There exists $\Delta > 0$, $\bar{F} > 0$ and $\epsilon' > 0$ such that

$$\sup_{0\le\theta\le\Delta} |f_\theta(t)| \le \bar{F}, \quad t \ge 0, \quad \text{and} \quad S(\epsilon) < +\infty \quad \epsilon > \epsilon'.$$

(B) Y is a.s. bounded.

Proof Note $Y(t) = Y_0(t) + y(t)$ for $t \ge 0$, where y is the solution of (14). For part (a), we prove first (A) implies (B). The first part of (A) gives $y(t) \to 0$ as $t \to \infty$, by the implication in Theorem 5 that (B) implies (A). The second part of (A) gives $Y_0(t) \to 0$ as $t \to \infty$ a.s., by part (a) of Theorem 12. Hence $Y(t) \to 0$ as $t \to \infty$.

To show (B) implies (A), note $Y(t)$ is Gaussian distributed for each t, with mean $y(t)$. Consider any sequence $t_n \uparrow \infty$. Then $Y_n := Y(t_n)$ converges to zero a.s., by (B). But since Y_n is Gaussian, it means that it also converges in distribution, which forces $\mathbb{E}[Y_n] \to 0$ as $n \to \infty$, or $y(t_n) \to 0$ as $n \to \infty$. Suppose now, by way of contradiction, that $\limsup_{t\to\infty} |y(t)| > 0$. Then there exists $c > 0$ and a sequence $t_n \uparrow \infty$ such that $|y(t_n)| > c$ for all n sufficiently large. But this contradicts the fact that (B) implies $y(t_n) \to 0$ as $n \to \infty$ for any divergent sequence (t_n). Hence, we must have $\limsup_{t\to\infty} |y(t)| = 0$, or $y(t) \to 0$ as $t \to \infty$. But by the implication in Theorem 5 that (A) implies (B), we must have that $\int_t^{t+\theta} f(s)\,ds \to 0$ as $t \to \infty$, which proves the first part of (A). To prove the second part, note we have shown $y(t) \to 0$ as $t \to \infty$, and assumed $Y(t) \to 0$ as $t \to \infty$ a.s. Hence $Y_0(t) = Y(t) - y(t) \to 0$ as $t \to \infty$ a.s. Now apply part (a) of Theorem 12 to see that we must have $S(\epsilon) < +\infty$ for all $\epsilon > 0$, which is the second part of (A). Thus we have shown that (B) implies (A), establishing the required equivalence to complete the proof of part (a).

For part (b), we prove first (A) implies (B). Note the first part of (A) implies y is bounded, applying Theorem 7 with $\gamma(t) \equiv 1$. The second part of (A) implies Y_0 is a.s. bounded, by part (b) of Theorem 12. Thus $Y = Y_0 + y$ is a.s. bounded, giving (B).

To show that (B) implies (A), it is enough to show that y is bounded; in that case the first part of (A) holds, by the converse half of Theorem 7 with $\gamma(t) \equiv 1$. On the other hand, if y is bounded, then $Y_0 = Y - y$ is a.s. bounded, by hypothesis. Thus, by part (b) of Theorem 12, the second part of (A) holds, and hence (B) implies (A). Hence (A) and (B) are equivalent, establishing part (b) of the theorem.

It remains to show that (B) implies y is bounded. Let m be any positive, non–decreasing continuous function with $m(t) \to \infty$ as $t \to \infty$ and (t_n) any sequence such that $t_n \uparrow \infty$ as $n \to \infty$. Now, $Y_n := Y(t_n)/m(t_n)$ is a sequence of Gaussian random variables with mean $y(t_n)/m(t_n)$, and $Y_n \to 0$ as $n \to \infty$ a.s. since Y is a.s. bounded. But then Y_n converge in distribution, so $y(t_n)/m(t_n) \to 0$ as $n \to \infty$.

Suppose next, by way of contradiction, that $\limsup_{t\to\infty} |y(t)| = +\infty$. Then $y^*(t) := \max_{0\le s\le t} |y(s)| \to \infty$ as $t \to \infty$. Pick $m(t) = \sqrt{y^*(t)}$, so that $m(t) \to \infty$ as $t \to \infty$, and let t_n be a sequence along which $y^*(t) = |y(t)|$. Then $|y(t_n)|/m(t_n) = y^*(t_n)/\sqrt{y^*(t_n)} = \sqrt{y^*(t_n)} \to \infty$ as $n \to \infty$, which contradicts what was shown at the end of the last paragraph. Hence the supposition $\limsup_{t\to\infty} |y(t)| = +\infty$ is false, and y is bounded, as required. □

This result, together with the discussion before Theorem 12, gives a characterisation of stability and boundedness for (23).

Theorem 14 *Suppose that $q \in (0, 1)$ and (2) holds. Let $f, \sigma \in C([0, \infty); \mathbb{R})$ and X and Y be the unique continuous adapted solutions to (23) and (26). Then*

(i) Condition (a)(A) of Theorem 13 $\iff X(t) \to 0$ as $t \to \infty$ a.s.;
(ii) Condition (b)(A) of Theorem 13 $\iff X$ is bounded a.s.

Some remarks are in order; first, Theorem 13 is to the best of our knowledge (and remarkably, given the simplicity of the equation) the first pathwise characterisation of asymptotic stability and boundedness for a stochastic differential equation which is subjected to both deterministic and stochastic perturbations. *We consider this an auxiliary result of tremendous importance*. We see straightaway that it can be used to characterise the a.s. convergence and boundedness of the stochastically and deterministically forced pantograph equation, an equation far more complicated that the SDE, revealing that *the stability conditions are independent of the structure of the underlying differential system.* Again, such a stability characterisation for deterministically and stochastically perturbed functional differential equations is, to the best of our knowledge, the first to appear in the literature.

To see the potential for Theorem 13 to have very wide applicability to characterise many other complicated differential systems which are forced state–independently, we give a sketch of the possible direction of research. Suppose, concretely, we have the following stochastic functional differential equation

$$dX(t) = (F(t, X_t) + f(t))\,dt + \sigma(t)\,dB(t), \quad t \geq 0,$$

where F is a functional on $[0, \infty) \times C([0, \infty))$. We view X as a perturbation of the underlying deterministic system $u'(t) = F(t, u_t)$. We use the common notational convention for functional differential equations that for a continuous function $x : \mathbb{R}^+ \to \mathbb{R}$, x_t refers to the history of x i.e., to $\{x(s) : 0 \leq s \leq t\}$. Now, consider Y as above, and let $Z := X - Y$. Then

$$Z'(t) = F(t, Z_t + Y_t) + Y(t), \quad t \geq 0.$$

Let F satisfy a uniform functional Lipschitz condition in the state i.e., if for all ϕ, $\psi \in C(\mathbb{R}^+; \mathbb{R})$, there exists a constant $K > 0$ such that

$$|F(t, \phi_t) - F(t, \psi_t)| \leq K \max_{0\leq s\leq t} |\phi(s) - \psi(s)|, \quad t \geq 0.$$

Note that this condition is sufficient to guarantee a unique continuous adapted solution of the original SFDE on $[0, \infty)$ (see e.g., Mao 2008). Thus, writing

$\tilde{f}(t) := F(t, Z_t + Y_t) - F(t, Z_t) + Y(t)$ for $t \geq 0$, we see that $\tilde{f}(t) \to 0$ as $t \to \infty$, $\tilde{f}$ is continuous on $[0, \infty)$, and Z obeys

$$Z'(t) = F(t, Z_t) + \tilde{f}(t), \quad t \geq 0.$$

Now, if a good perturbation theory exists for $C_0(\mathbb{R}^+; \mathbb{R})$ perturbations of the equation $u'(t) = F(t, u_t)$, we may be able to show that $Z(t) \to 0$ as $t \to \infty$, and hence that $X(t) = Y(t) + Z(t) \to 0$ as $t \to \infty$.

In the converse direction, stability of the unperturbed functional differential equation means that we should impose the condition

$$F(t, \phi_t) \to 0 \quad \text{as } t \to \infty \text{ for each } \phi \in C_0(\mathbb{R}^+; \mathbb{R}).$$

By our usual arguments, Y can be connected to X via

$$Y(t) = X(t) - X(0)e^{-t} - \int_0^t e^{-(t-s)}\{X(s) + F(s, X_s)\}\, ds, \quad t \geq 0.$$

Therefore, if $X(t) \to 0$ as $t \to \infty$, the term in curly brackets in the integrand is a continuous function in s, and tends to 0 as $s \to \infty$. Therefore, the integral tends to zero as $t \to \infty$, and hence $Y(t) \to 0$ as $t \to \infty$. From this brief discussion, it can be seen how an effective convergence theory for many "doubly perturbed" functional differential equations can be developed. In rough terms, provided the asymptotic stability of the unperturbed is robust to deterministic C_0 perturbations, then the convergence of the stochastic FDE system is equivalent to the convergence of the SDE, which has already been characterised in Theorem 13.

6 Developments and Further Work

The principal purpose of this paper was to develop necessary and sufficient conditions for the convergence of perturbed solutions of a celebrated unbounded delay differential equation to the equilibrium solution of the underlying unforced equation. Making these very particular choices, we have demonstrated that the forcing terms which are necessary and sufficient to preserve the asymptotic stability are precisely those needed to give the convergence to equilibrium of simple linear differential equations (whether deterministic or stochastic).

One reason to focus on the pantograph equation, rather than to consider more general equations, is to exploit the fact that the asymptotic behaviour of the unperturbed pantograph equation is (essentially) completely understood, whereas for equations with general unbounded delay, and finite dimensional equations, for example, complete characterisations of the asymptotic behaviour are not always available. This makes it harder to determine whether gaps in the analysis for perturbed equations are purely the result of shortcomings in the perturbed (or stochastic) analysis, or may instead reflect the incomplete state of knowledge in the unperturbed systems. The abundance of necessary and sufficient conditions for various phenomena in the preceding questions supports this conservative approach.

On the other hand, pantograph equations represent only a particular example of equations with unbounded delay, and it can be the case that the arguments used to analyse scalar equations can possess special features which make generalisation to higher dimensions difficult, or even impossible. Therefore, in this section, we give arguments to show that a number of generalisations are possible, despite possibly incomplete knowledge about underlying unperturbed equations, or other important auxiliary equations (e.g., the SDE satisfied by Y). Leaving aside questions about growth or decay rates for general problems momentarily, several avenues for generalisation may be readily stated:

(i) Equations with general unbounded delay, rather than simply proportional delay;
(ii) Multi–dimensional equations;
(iii) Equations with multiplicative, rather than additive noise e.g., equations of the form
$$dX(t) = (bX(t) + aX(qt))\,dt + \sigma X(t)\,dB(t), \quad t \geq 0.$$

In the following subsections, we sketch details of how the results in the earlier part of the paper can be generalised (rather successfully) to cover these cases. In cases (i) and (ii), the machinery developed in earlier sections can mostly be generalised in reasonably obvious ways (with some new problems emerging), while in case (iii), although the approach of the earlier part of the paper is not useful in the particulars, it nevertheless suggests an analogous approach to attacking the equations in question which seems to generate some interesting preliminary results.

6.1 Equations with General Unbounded Delay

For the first question, the framework of the present work is sufficient to provide a positive answer. Consider now the SFDE

$$dX(t) = (bX(t) + aX(t - \tau(t))) + f(t))\,dt + \sigma(t)\,dB(t), \quad t \geq 0, \tag{32}$$

where τ is a positive continuous function with $\sup_{t\geq 0} t - \tau(t) =: -\bar{\tau} > -\infty$ and $t - \tau(t) \to \infty$ as $t \to \infty$. In this case, we need an initial function $\psi \in C([-\bar{\tau}, \infty); \mathbb{R})$ so that if $X = \psi$ on $[-\bar{\tau}, 0]$, then there will exist a unique continuous adapted solution of the equation on $[-\bar{\tau}, \infty)$. Confining attention to the situation when $X(t) \to 0$ as $t \to \infty$ a.s., we may proceed as above to show that if $X(t) \to 0$ as $t \to \infty$ a.s., then $Y(t) \to 0$ as $t \to \infty$ a.s. We do this by writing $dX(t) = \{-X(t) + (1 + b)X(t) + aX(t - \tau(t)) + f(t)\}\,dt + \sigma(t)\,dB(t)$, and now, by treating everything apart from

$-X(t)$ on the right hand side as a perturbation, using stochastic integration by parts, and rearranging for Y, we get

$$Y(t) = X(t) - e^{-t}X(0) - \int_0^t e^{-(t-s)}\{(1+b)X(s) + aX(s-\tau(s))\}\,ds, \quad t \geq 0. \tag{33}$$

Since $t - \tau(t) \to \infty$ as $t \to \infty$, the term in curly brackets tends to zero as $s \to \infty$, so the integral tends to zero, and thus $Y(t) \to 0$ as $t \to \infty$ a.s., as claimed.

To show $Y(t) \to 0$ as $t \to \infty$ implies $X(t) \to 0$ as $t \to \infty$ a.s., we may define $Z(t) = X(t) - Y(t)$ for $t \geq -\bar{\tau}$ (extending Y to be zero on $[-\bar{\tau}, 0]$), in which case $Z = \psi$ on $[-\bar{\tau}, 0]$ and Z obeys

$$Z'(t) = bZ(t) + aZ(t - \tau(t)) + \phi(t), \quad t \geq 0$$

where $\phi(t) = (1+b)Y(t) + aY(t - \tau(t))$ for $t \geq 0$. Clearly, if $Y(t) \to 0$ as $t \to \infty$ a.s., then $\phi(t) \to 0$ as $t \to \infty$ a.s. Evidently, if $Y(t) \to 0$ as $t \to \infty$ a.s., then $Z(t) \to 0$ as $t \to \infty$ a.s. is a necessary and sufficient condition to ensure that $X(t) \to 0$ as $t \to \infty$ a.s.

Thus, under the condition that the perturbations f and σ are "small" in such a way that the solutions of the SDE converge to zero, the question of whether deterministic and stochastic perturbations which preserve the convergence of solutions to the equilibrium in the complicated SFDE reduces to whether the deterministically perturbed FDE is convergent if there is a fading perturbation. By virtue of Čermák (2002), it is known that the conditions $b < 0$ and $|a| < |b|$ are sufficient for the solution of the unperturbed equation

$$u'(t) = bu(t) + au(t - \tau(t)), \quad t \geq 0,$$

to obey $u(t) \to 0$ as $t \to \infty$. Keeping this condition on a and b, we can take the equation for Z and show that Z obeys

$$D_+|Z(t)| \leq b|Z(t)| + |a||Z(t - \tau(t))| + |\phi(t)|, \quad t \geq 0,$$

where D_+ is the Dini–derivative earlier introduced. Now, by comparison arguments, $|Z(t)| < z(t)$ for $t \geq -\tau$ where $z = |\psi| + 1$ on $[-\tau, 0]$ and

$$z'(t) = bz(t) + |a|z(t - \tau(t)) + |\phi(t)| + e^{-t}, \quad t \geq 0.$$

See Appleby and Buckwar (2010), for example. Finally, it remains to show that $z(t) \to 0$ as $t \to \infty$, and this will give $Z(t) \to 0$ as $t \to \infty$, as needed.

The proof that $z(t) \to 0$ as $t \to \infty$ can be achieved by means of comparison arguments, showing first that z is bounded, and then by showing that a positive limsup of z leads to a contradiction, and hence to the conclusion that $z(t) \to 0$ as $t \to \infty$.

The boundedness of z may also be proven by a comparison argument: simply choose a (sufficiently large) constant upper solution to the differential equation for z, exploiting the facts that the last terms on the right hand side are uniformly bounded, and that $-b > |a| \geq 0$. The key to the proof of convergence is to view all the terms in z, apart from the instantaneous term, as perturbations of the differential equation $v' = bv$, writing down the variation of constants formula, majorising, and then taking the limsup as $t \to \infty$. From this, it is seen that a positive limsup for z is incompatible with the inequality $-b > |a| \geq 0$.

Parallel arguments can be developed for the characterisation of bounded solutions. Thus, the following theorem can be established.

Theorem 15 *Let* $b < 0$, $|a| < |b|$, $\tau \in C([0,\infty);(0,\infty))$ *be such that* $t - \tau(t) \to \infty$ *as* $t \to \infty$ *and* $-\bar{\tau} := \sup_{t\geq 0}(t - \tau(t)) > -\infty$. *Let* f *and* σ *be continuous, and suppose that* $\psi \in C([-\bar{\tau}, 0]; \mathbb{R})$. *Then there is a unique continuous adapted process* X *which is the solution of* (32) *with with* $X = \psi$ *on* $[-\bar{\tau}, 0]$, *and moreover:*

(a) The following are equivalent:

(A) f *and* σ *obey condition (a)(A) of Theorem 13;*
(B) $X(t) \to 0$ *as* $t \to \infty$ *a.s.*

(b) The following are equivalent:

(A) f *and* σ *obey condition (b)(A) of Theorem 13;*
(B) X *is a.s. bounded.*

We note that this theorem does not require any smoothness or monotonicity properties on the delay τ, nor on the delayed argument id $- \tau$. This improves upon the regularity required on τ in Čermák (2002), for instance.

6.2 Decay Estimates for General Unbounded Delay Equations

The question about rates of decay to zero if the solution of (32) demands that a good asymptotic estimate be found to the rate at which $Y(t) \to 0$ as $t \to \infty$. This can then be used to get the rate of decay of $\phi(t) \to 0$ as $t \to \infty$. At this point, it may be hard to use (Čermák 2002, Theorem 2.3) on perturbed equations, since the asymptotic behaviour of ϕ' is also needed, and ϕ' will not be well–defined for stochastic perturbations, because the non–differentiability of Y will make ϕ non–differentiable. There are possible routes to circumvent this problem. One is to invoke a kind of comparison argument in the manner of Appleby and Buckwar (2010) or this paper, but this may not lead to sharp results. Another possibility is to apply a "double mollification" to the equation, as follows. Consider X, Y and Z as above. Now, define $Y_2'(t) = bY_2(t) + \phi(t)$ for $t \geq 0$, and set $Y_2(t) = 0$ for $t \leq 0$. Since

$b < 0$ and $\phi(t) \to 0$ as $t \to \infty$, we have that $Y_2(t) \to 0$ as $t \to \infty$. Moreover, an upper estimate on the rate of decay of Y_2 can be obtained from the rate of decay of ϕ (which is known from Y). Next, define $Z_2 = Z - Y_2$. Then

$$Z_2'(t) = bZ_2(t) + aZ_2(t - \tau(t))) + \phi_2(t)$$

where $\phi_2(t) = aY_2(t - \tau(t))$. Now, provided τ is continuously differentiable, as it is in Cermak's theory, ϕ_2 is continuously differentiable, and asymptotic estimates can be obtained for ϕ_2 and ϕ_2' in terms of Y_2 and $Y_2' = bY_2 + \phi$, all of which can be estimated using pathwise estimates of Y. At this point, Cermak's results could be applied off the shelf to give an estimate on Z_2. But since $Z = Y_2 + Z_2$ and $X = Z + Y$, an estimate on the decay rate of X could be found, again appealing to the decay estimate on Y. As a result, we see that a theory that gives good upper bounds for the rate of decay of X can probably be deduced from existing results, provided we can establish good asymptotic estimates for the solution of the SDE. Such a characterisation of the asymptotic behaviour of the SDE will form part of a future work.

The sharpness of the sufficient conditions generated in this approach could be examined by returning to (33). If X has a known a.s. rate of decay, by estimating the terms on the righthand side, an estimate on the necessary rate of decay of Y could be found.

6.3 *Multidimensional Equations with Additive Noise*

In this subsection, we discuss the multidimensional pantograph equation (rather than equations with general unbounded delay) which is forced both deterministically and stochastically (although, as we mention at the end, most of our argument remains valid for equations with general unbounded delay). We consider d–dimensional equations. The analogue of the scalar deterministic perturbation is an $f \in C([0, \infty); \mathbb{R}^d)$ (and we denote its i-th component f_i). The multidimensional analogue for deterministically perturbed scalar equations is therefore

$$x'(t) = Bx(t) + Ax(qt) + f(t), \quad t \geq 0,$$

where A and B are $d \times d$ real matrices, based on the underlying unforced equation being

$$u'(t) = Bu(t) + Au(qt), \quad t \geq 0. \tag{34}$$

We assume hereinafter that A is not the zero matrix, since in that case all delay differential equations reduce to being ordinary ones.

As for the stochastic equation, since there are many dimensions, there may reasonably be many (independent) sources of randomness affecting the dynamics. Focussing on the dynamics of the ith component of the solution vector $X(t)$, if we are to have a state–independent additive noise, we arrive at:

$$dX_i(t) = \{[BX(t) + AX(qt)]_i + f_i(t)\}\, dt + \sum_{j=1}^{r} \sigma_{ij}(t)\, dW_j(t),$$

where $W = (W_1, \ldots, W_r)$ is an r–dimensional standard Brownian motion (or Wiener process). The $d \times r$ scalar continuous functions σ_{ij} can be formed into a $d \times r$ continuous $d \times r$ matrix–values function (its value at time t is denoted $\Sigma(t)$), and then X can be expressed in the differential shorthand

$$dX(t) = \{BX(t) + AX(qt) + f(t)\}\, dt + \Sigma(t)\, dW(t), \quad t \geq 0. \tag{35}$$

With $X(0) = \xi \in \mathbb{R}^d$ given, there is a unique continuous adapted process satisfying this equation. The question now arises: can we characterise the situation wherein $x(t) \to 0$ as $t \to \infty$ and $X(t) \to 0$ as $t \to \infty$, granted that $u(t) \to 0$ as $t \to \infty$?

As before, we start with a converse result. Let Y be the d–dimensional process given by

$$dY(t) = \{-Y(t) + f(t)\}\, dt + \Sigma(t)\, dW(t), \quad t \geq 0,$$

with $Y(0) = 0 \in \mathbb{R}^d$. Then generalising existing arguments to the multidimensional case, we get

$$Y(t) = X(t) - e^{-t}X(0) - \int_0^t e^{-(t-s)}\{(I_d + B)X(s) + AX(qs)\}\, ds, \quad t \geq 0,$$

where I_d is the $d \times d$ identity matrix. From this, we see that whenever $X(t) \to 0$ as $t \to \infty$ a.s., we have $Y(t) \to 0$ as $t \to \infty$ a.s. This last condition holds provided $Y_i(t) \to 0$ as $t \to \infty$ for each i a.s., and noticing that Y_i obeys the scalar SDE

$$dY_i(t) = \{-Y_i(t) + f_i(t)\}\, dt + \sum_{j=1}^{r} \sigma_{ij}(t)\, dW_j(t),$$

we see that the equation satisfied by Y_i is of essentially the same form as that satisfied by the solution of (26). In fact, if we define

$$\sigma_i(t) := \left(\sum_{j=1}^{r} \sigma_{ij}^2(t) \right)^{1/2}, \quad t \geq 0, \tag{36}$$

we may use the martingale representation theorem (see e.g., Karatzas and Shreve 1991) to rewrite the continuous martingale

$$M_i(t) := \sum_{j=1}^{r} \int_0^t \sigma_{ij}(s)\,dW_j(s), \quad t \ge 0,$$

as

$$M_i(t) = \int_0^t \sigma_i(s)\,d\tilde{B}_i(s), \quad t \ge 0,$$

where each $\tilde{B}_i$ is a standard one–dimensional Brownian motion (see e.g., Karatzas and Shreve 1991). Therefore, Y_i obeys the SDE

$$dY_i(t) = \{-Y_i(t) + f_i(t)\}\,dt + \sigma_i(t)\,d\tilde{B}_i(t),$$

which is in precisely the form of (26). Now, define $S_i(\epsilon)$ as $S(\epsilon)$ in (3), but with σ_i in the role of σ. Since we require $Y_i(t) \to 0$ as $t \to \infty$ for each i, a.s., it follows from the converse result (i.e. (B) implies (A)) of part (a) of Theorem 13, applied to each Y_i, that $S_i(\epsilon) < +\infty$ for all $\epsilon > 0$ and that for each f_i and each $\theta > 0$ the condition

$$\int_{t-\theta}^{t} f_i(s)\,ds \to 0 \quad t \to \infty$$

holds. This latter condition is nothing but

$$\int_{t-\theta}^{t} f(s)\,ds \to 0, \quad t \to \infty, \quad \text{for all } \theta > 0. \tag{37}$$

Therefore this condition and $S_i(\epsilon) < +\infty$ for all $\epsilon > 0$ and $i \in \{1, \ldots, d\}$ is *necessary* in order that $X(t) \to 0$ as $t \to \infty$ a.s.

Conversely, if each $S_i(\epsilon) < +\infty$ for all $\epsilon > 0$ and $i \in \{1, \ldots, d\}$ and $\int_{t-\theta}^{t} f(s)\,ds \to 0$ as $t \to \infty$, then $Y_i(t) \to 0$ as $t \to \infty$ a.s. for each i, and therefore $Y(t) \to 0$ as $t \to \infty$ a.s. As before, set $Z(t) := X(t) - Y(t)$ for $t \ge 0$, and let $\phi(t) = (I_d + B)Y(t) + AY(qt)$ for $t \ge 0$. Then $Z(0) = X(0)$ and Z obeys the perturbed equation

$$Z'(t) = BZ(t) + AZ(qt) + \phi(t), \quad t \ge 0. \tag{38}$$

Moreover, since $Y(t) \to 0$ as $t \to \infty$ a.s., we have that $\phi(t) \to 0$ as $t \to \infty$ a.s., and since the paths of Y are continuous, ϕ is continuous on $[0, \infty)$. We notice that if $Z(t) \to 0$ as $t \to \infty$ a.s., then $X(t) \to 0$ as $t \to \infty$ a.s.

Therefore, by considering (38) path–by–path (i.e., for each ω in the a.s. event where X and Y are well–defined, and Y tends to zero), it can be seen that if the *deterministic* property

$$\text{For every } \phi \in C_0([0,\infty);\mathbb{R}^d)\text{, the solution } Z \text{ of (38) obeys } Z(t) \to 0 \text{ as } t \to \infty \tag{39}$$

holds, then we have that $Y(t) \to 0$ as $t \to \infty$ a.s. implies $X(t) \to 0$ as $t \to \infty$ a.s.

We have therefore proven the following result, *which eliminates the need for further analysis of stochastically perturbed equations*, and reduces the study of (35) to that of (38) with deterministic forcing term $\phi \in C_0([0,\infty);\mathbb{R}^d)$.

Theorem 16 *Let $f \in C([0,\infty);\mathbb{R}^d)$ and $\Sigma \in C([0,\infty);\mathbb{R}^{d\times r})$. With $\sigma_{ij} = \Sigma_{ij}$, define σ_i for each $i \in \{1,\dots,d\}$ according to* (36). *Suppose that the property* (39) *holds for solutions of* (38). *Let X be the unique continuous adapted process which obeys* (35). *The following are then equivalent:*

(A) f obeys (37) *and for each i and $\epsilon > 0$*

$$S_i(\epsilon) := \sum_{n=1}^{\infty} \sqrt{\int_{n-1}^{n} \sigma_i^2(s)\,ds}\,\exp\left(-\epsilon \int_{n-1}^{n} \sigma_i^2(s)\,ds\right) < +\infty. \tag{40}$$

(B) $X(t) \to 0$ as $t \to \infty$ a.s.

Of course, the theorem would be vacuous if property (39) does not hold. We now explore the property (39) a little further, and show that it holds under conditions which are, in a certain sense, reasonable multidimensional analogues of (2). Since (39) must hold in the case $\phi(t) = 0$ for all $t \geq 0$, this means the all solution u of the unperturbed pantograph equation (34) must obey $u(t) \to 0$ as $t \to \infty$. Iserles has shown in Iserles (1993) that this is equivalent to

$$\text{All eigenvalues of } B \text{ have negative real parts} \tag{41}$$

(a condition which implies B is invertible), together with the spectral radius condition

$$\rho(B^{-1}A) < 1. \tag{42}$$

Therefore, the conditions (41) and (42) should be imposed on A and B. However, to the best of our knowledge, no result exists in the literature which states that (41) and (42) are sufficient to ensure that property (39) holds. It would clearly be of value to prove such a result, since, if it holds, a complete characterisation of convergence of solutions of perturbed stochastic and deterministic pantograph equations would be achieved. A cursory reading of the work of Iserles in Iserles (1993), however, does not lead to optimism that his approach could fulfill this easily. The proofs in that paper rely on analyticity of the solutions of the unperturbed equation, and so–called

Dirichlet series solutions that can be formed as a consequence of the analyticity. For stochastic equations at least, the forcing function ϕ is continuous, but nowhere differentiable, so Z is a C^1 function, but not even C^2, for instance.

We show next, however, that for a subclass of A and B obeying properties (41) and (42), that property (39) does hold. We retain property (41), and note that (42) in a certain sense demands that A be small relative to B. Our sufficient condition imposes such a smallness condition on A in lieu of (42).

It is worth noting that although our argument is not sharp in the pantograph case, it works without modification for general delay; this is appealing, because, to the best of our knowledge, necessary and sufficient conditions such as (41) and (42) are not known which characterise the stability of unperturbed multidimensional equations with unbounded delay.

To prove property (39), we continue to assume (41); therefore it follows that there is a unique symmetric, positive definite $d \times d$ matrix Q (with real entries) such that

$$B^T Q + QB = -I_d, \tag{43}$$

where, here and in the sequel, the superscripted T denotes the transpose of a matrix. Now consider $z(t) := Z(t)^T QZ(t)$ for $t \geq 0$. We see that $z(t) \in \mathbb{R}$ for each $t \geq 0$. Furthermore, since Q is positive definite, $z(t) \geq 0$ for all $t \geq 0$. Our strategy is to derive a perturbed scalar pantograph differential inequality for z. Using (43) we get

$$\begin{aligned} z'(t) &= (Z(t)^T B^T + Z(qt)^T A^T + \phi(t)^T)QZ(t) \\ &\quad + Z(t)^T(QBZ(t) + QAZ(qt) + Q\phi(t)) \\ &= -\|Z(t)\|^2 + 2Z(t)^T QAZ(qt) + 2\phi(t)^T QZ(t), \quad t \geq 0. \end{aligned} \tag{44}$$

Notice next that the positive definiteness of Q means that $c_1\|Z(t)\|^2 \leq z(t) \leq c_2\|Z(t)\|^2$ for some $0 < c_1 \leq c_2$, where $\|\cdot\|$ is the Euclidean norm on $\mathbb{R}^d$ (in fact, c_1 and c_2 are the minimum and maximum eigenvalues of Q, respectively).

Now we estimate the terms on the righthand side of (44). We start with the last term. By the Cauchy–Schwarz inequality, we get

$$|2\phi(t)^T QZ(t)| \leq 2\|\phi(t)\|\|QZ(t)\| \leq \|\phi(t)\|\|Q\|\|Z(t)\| =: 2\epsilon_1(t)\|Z(t)\|,$$

where $\epsilon_1(t) := \|\phi(t)\|\|Q\|$ is continuous, nonnegative and tends to zero as $t \to \infty$, and for any $d \times d$ matrix M, $\|M\|$ denotes the matrix norm of M induced from the Euclidean norm. Now, recall that $2xy \leq x^2 + y^2$ for all $x, y \geq 0$, so, putting $x = \sqrt{\epsilon_1(t)}$ and $y = \sqrt{\epsilon_1(t)}\|Z(t)\|$, we get

$$|2\phi(t)^T QZ(t)| \leq 2\epsilon_1(t)\|Z(t)\| \leq \epsilon_1(t) + \epsilon_1(t)\|Z(t)\|^2. \tag{45}$$

For the second term on the righthand side of (44), we have $2Z(t)^T QAZ(qt) = 2Z(t)^T Q^T AZ(qt) = 2(A^T QZ(t))^T Z(qt)$, so, setting $\beta := \|A^T Q\| > 0$ and using the Cauchy–Schwarz inequality once again, we have

$$\begin{aligned}|2Z(t)^T QAZ(qt)| &\leq 2\|A^T QZ(t)\|\|Z(qt)\| \leq 2\beta\|Z(t)\|\|Z(qt)\| \\ &\leq \beta\left(\alpha\|Z(t)\|^2 + \frac{1}{\alpha}\|Z(qt)\|^2\right),\end{aligned}$$

where, for arbitrary $\alpha > 0$ we have used the inequality $2xy \leq \alpha x^2 + y^2/\alpha$ for $x, y \geq 0$ at the last step.

Hence, using this estimate and (45), we may return to (44) to get

$$z'(t) \leq -\|Z(t)\|^2 + \beta\alpha\|Z(t)\|^2 + \frac{\beta}{\alpha}\|Z(qt)\|^2 + \epsilon_1(t)\|Z(t)\|^2 + \epsilon_1(t).$$

Next, we note that $\beta > 0$; to see this, assume to the contrary that $\beta = 0$. Then $A^T Q = 0$. Since Q is positive definite, Q is invertible. This forces $A = 0$, which we have ruled out by hypothesis, and we have the desired contradiction. Thus, we have that $\beta > 0$ and we may pick $\alpha := 1/(2\beta) > 0$. Then

$$z'(t) \leq -\frac{1}{2}\|Z(t)\|^2 + 2\beta^2\|Z(qt)\|^2 + \epsilon_1(t)\|Z(t)\|^2 + \epsilon_1(t), \quad t \geq 0.$$

Since $\epsilon_1(t) \to 0$ as $t \to \infty$, for every $\epsilon \in (0, 1/2)$ there is $T(\epsilon) > 0$ such that for $t \geq T(\epsilon)$, $\epsilon_1(t) < \epsilon$. Hence for $t \geq T(\epsilon)$, we have

$$z'(t) \leq -\left(\frac{1}{2} - \epsilon\right)\|Z(t)\|^2 + 2\beta^2\|Z(qt)\|^2 + \epsilon_1(t).$$

Finally, $\|Z(qt)\|^2 \leq c_1^{-1}z(qt)$ and $\|Z(t)\|^2 \geq c_2^{-1}z(t)$, so for $\epsilon \in (0, 1/2)$ we have

$$z'(t) \leq -\left(\frac{1}{2} - \epsilon\right)\frac{1}{c_2}z(t) + \frac{2\beta^2}{c_1}z(qt) + \epsilon_1(t), \quad t \geq T(\epsilon). \tag{46}$$

Since for any $d \times d$ matrix M the identity $\|M\| = \|M^T\|$ holds for the 2–norm, and Q is symmetric, we have $\beta = \|Q^T A\| = \|QA\|$. Now, suppose that

$$4\|QA\|^2 < \frac{c_1}{c_2}, \tag{47}$$

or equivalently, $2\beta^2 c_2/c_1 < 1/2$. Choose $\epsilon \in (0, 1/2 - 2\beta^2 c_2/c_1)$. Then the coefficient of $z(t)$ in (46) is negative, and in modulus exceeds the coefficient of $z(qt)$. Since $\epsilon_1(t) \to 0$ as $t \to \infty$, we may apply our scalar theory to the inequality (46) to deduce that $z(t) \to 0$ as $t \to \infty$. But since $\|Z(t)\|^2 \leq c_1^{-1}z(t)$, this implies $Z(t) \to 0$ as $t \to \infty$.

Thus we see the sufficient conditions required to ensure property (39) are (41) together with (47) in place of (42). We discuss now the condition (47), and show that it retains much of the spirit of the condition (42).

The righthand side of (47) is the minimal eigenvalue of Q, divided by the maximal eigenvalue of Q, so is a quantity which depends on B, but not on A. Therefore, as earlier claimed, (47) can be interpreted as a smallness condition on A, similar to (42), because, using the submultiplicative property of the matrix 2–norm, and the fact that $\|Q\| = c_2$ for this norm, we see that

$$4\|A\|^2 < \frac{c_1}{c_2^3}$$

implies (47). Note also that the condition (47) is "delay independent" in the sense that (47) does not depend on q; it also shares this property with the condition (42).

Using the fact that $Q = \int_0^\infty e^{Bs} e^{B^T s}\, ds$ and $\int_0^\infty e^{2Bs}\, ds = -\frac{1}{2}B^{-1}$, we see that we can rewrite (42) and (47) as

$$\rho\left(\int_0^\infty e^{2Bs}\, ds A\right) < \frac{1}{2}, \quad \left\|\int_0^\infty e^{Bs} e^{B^T s}\, ds A\right\| < \frac{1}{2}\sqrt{\frac{\text{minimal eigenvalue of } Q}{\text{maximal eigenvalue of } Q}},$$

respectively. Noting that $\rho(M) \leq \|M\|$ for any matrix M, we see the close connection between the conditions.

To see further similarities between (47) and (42), note in the one–dimensional case that we $b := B < 0$ and $2BQ = -1$. Since $c_1/c_2 = 1$, the condition (47) is equivalent to $1 = \sqrt{c_1/c_2} > 2\|QA\| = 2QA = -A/B = -a/b$, and hence in the case $d = 1$, (41) and (47) are exactly (2), which is exactly (41) and (42) in the case $d = 1$.

One other connection between (42) and (47) is worth pointing out. If B is real diagonal, then $Q = -\frac{1}{2}B^{-1}$. Suppose A is symmetric. Then $QA = B^{-1}A$ is symmetric, and so $\|QA\| = \rho(QA) = \rho(B^{-1}A)$ and (47) becomes

$$\rho(B^{-1}A) < \frac{\text{minimal eigenvalue of } |B|}{\text{maximal eigenvalue of } |B|},$$

a condition very near in form to, but more restrictive than, condition (42).

The above calculations therefore lead to the following theorem, which is a corollary of Theorem 16.

Theorem 17 *Let $f \in C([0,\infty); \mathbb{R}^d)$ and $\Sigma \in C([0,\infty); \mathbb{R}^{d \times r})$. With $\sigma_{ij} = \Sigma_{ij}$, define σ_i for each $i \in \{1, \dots, d\}$ according to* (36). *Suppose that* (41) *holds, and with Q defined by* (43), *the condition* (47) *holds. Let X be the unique continuous adapted process which obeys* (35). *Then the following are equivalent:*

(A) f *obeys* (37) *and for each* i *and* $\epsilon > 0$

$$S_i(\epsilon) := \sum_{n=1}^{\infty} \sqrt{\int_{n-1}^{n} \sigma_i^2(s)\,ds}\, \exp\left(-\epsilon \int_{n-1}^{n} \sigma_i^2(s)\,ds\right) < +\infty. \tag{48}$$

(B) $X(t) \to 0$ *as* $t \to \infty$ *a.s.*

As stated above, we believe an optimal result would replace the sufficient condition (47) with (42).

We finish with the remarks that the sketched results in this section do not really depend on proportional delay, and therefore should hold for any multidimensional equation with unbounded delay, and also that a result characterising boundedness can be formulated and proved with equal ease.

6.4 Equations with State–Dependent Noise

There are many studies already present concerning the asymptotic behaviour of stochastic pantograph equations with multiplicative noise, or more generally, state–dependent noise. To date, the results on stability, growth bounds etc. tend to give upper estimates, and sufficient conditions for asymptotic stability. The question arises, therefore, whether the techniques of this paper might be helpful in obtaining more precise asymptotic results.

A direct application of the technique is unlikely to be successful in dealing with equations of the form

$$dX(t) = (bX(t) + aX(qt))\,dt + \sigma X(t)\,dB(t) \tag{49}$$

since there is no obvious small perturbation to mollify. However, part of the sharpness achieved in this paper is due to a successful decomposition of non–differentiable dynamics of the delay–differential equation into an ordinary non–smooth stochastic differential equation, and a pathwise differentiable stochastic delay differential equation, so an adoption of the that aspect of the philosophy of this paper may be useful. We show briefly here how, in the special case that $a > 0$, solutions of (49) have the following asymptotic behaviour.

Theorem 18 *Define* $\lambda = b - \sigma^2/2$. *Let* $a > 0$ *and* $X(0) > 0$. *Let* X *be the unique strong (positive) solution of* (49).

(a) If $\lambda > 0$, *then*

$$\lim_{t\to\infty} \frac{1}{t} \log X(t) = \lambda, \quad \text{a.s.}$$

(b) If $\lambda \leq 0$, then

$$\lim_{t\to\infty} \frac{1}{t} \log X(t) = 0, \quad a.s.$$

Therefore, the situation is similar to the case for the deterministic pantograph equation (wherein $\sigma = 0$), with subexponential asymptotic behaviour when the underlying instantaneous equation

$$dv(t) = bv(t)\,dt + \sigma v(t)\,dB(t)$$

is asymptotically stable (which occurs when $\lambda < 0$). When the underlying instantaneous equation is unstable however (when $\lambda > 0$), exponential growth occurs, and does so at a rate $\lambda = b - \sigma^2/2 > 0$.

This result can be established by means of the following multiplicative decomposition. Let ρ be given by $\rho(0) = 1$ and

$$d\rho(t) = b\rho(t)\,dt + \sigma\rho(t)\,dB(t), \quad t \geq 0.$$

Then ρ, like Y, is the solution of a linear SDE, and has an explicit solution

$$\rho(t) = e^{(b-\sigma^2/2)t+\sigma B(t)}, \quad t \geq 0.$$

Since ρ is strictly positive, $1/\rho$ is a well–defined process and obeys the SDE

$$d\left(\frac{1}{\rho(t)}\right) = \frac{\sigma^2 - b}{\rho(t)}\,dt - \frac{\sigma}{\rho(t)}\,dB(t), \quad t \geq 0.$$

The process $Z = X/\rho$ is likewise well–defined, and by stochastic integration by parts obeys

$$dZ(t) = X(t)d\left(\frac{1}{\rho(t)}\right) + \frac{1}{\rho(t)}dX(t) + \sigma X(t) \cdot -\frac{\sigma}{\rho(t)}\,dt,$$

which simplifies to

$$dZ(t) = \frac{1}{\rho(t)}aX(qt)\,dt, \quad t \geq 0.$$

Therefore, we have

$$Z(t) = X(0) + \int_0^t aZ(qs)\frac{\rho(qs)}{\rho(s)}\,ds, \quad t \geq 0.$$

Since the integrand is continuous, Z is in fact differentiable, and we have

$$Z'(t) = a\frac{\rho(qt)}{\rho(t)}Z(qt) =: a(t)Z(qt), \quad t \geq 0, \tag{50}$$

where $\lambda = b - \sigma^2/2$ and $a(t) = ae^{\lambda(q-1)t+\sigma(B(qt)-B(t))}$. The strategy now is to study the dynamics of Z pathwise, exploiting the fact that $a(t)$ (or equally ρ) is known explicitly in terms of the Brownian motion B. Finally, we can recover information about X from $X(t) = \rho(t)Z(t)$, again exploiting knowledge of ρ.

When $\lambda > 0$, this technique will work well, since in that case, due to the Strong Law of Large Numbers for standard Brownian motion,

$$\lim_{t\to\infty} \frac{1}{t} \log |a(t)| = -\lambda(1-q) < 0, \quad \text{a.s.},$$

so $a \in L^1(0,\infty)$ a.s. Then, arguing pathwise, Z tends to a finite limit a.s. (call it Z^*). Therefore

$$X(t) \sim Z^*\rho(t) = Z^*e^{\lambda t+\sigma B(t)}, \quad t \to \infty, \quad \text{a.s.},$$

so X grows exponentially (under the assumption that the random limit Z^* is non–trivial). We have $Z^* > 0$ in the case that $X(0) > 0$ and $a > 0$, where $Z(t) \uparrow Z^* \in (0,\infty)$ as $t \to \infty$.

In the case when $Z^* = 0$, Z tends to limit Z^* according to $\limsup_{t\to\infty} \log|Z(t)|/t \leq -\lambda$. This can be proved by obtaining an a priori exponential rate of decay on Z from (50), and then recursively estimating the rate of decay by feeding the exponential estimate back into (50). From the final estimate on Z, we end up with $\limsup_{t\to\infty} \log|X(t)|/t \leq 0$.

Therefore, in the general case (without sign restrictions on a and $X(0)$), we have a rather complete picture of the long–run dynamics when $\lambda > 0$. Either $Z^* \neq 0$ and X inherits the asymptotic behaviour of the ρ (and does so rapidly, because Z converges exponentially fast to Z^*) or $Z^* = 0$, and X grows more slowly than any exponential function.

The case $\lambda < 0$ is more analogous to the situation covered in this paper, since when $\sigma = 0$, $\lambda = b < 0$. In that situation, $a(t)$ still has exponential asymptotic behaviour, and for every $\epsilon \in (0, 1-q)$ there is a random variable $T(\epsilon) > 0$ such that

$$D_+|Z(t)| \leq \left(\frac{-\lambda(1-q-\epsilon)}{1-q} + \epsilon\right)e^{-\lambda(1-q-\epsilon)t}|Z(qt)|, \quad t \geq T(\epsilon).$$

Now, using comparison techniques, one can show by choosing $\bar{Z}$ sufficiently large (take $\bar{Z} > \max_{s\in[qT,T]} |Z(s)|$) that $|Z(t)| \leq \bar{Z}e^{-\lambda(1-q-\epsilon)/(1-q)t} =: Z_+(t)$ for all $t \geq T$. This is done by showing the Z_+ obeys the upper differential inequality satisfied by $|Z|$. From this we rapidly deduce that

$$\limsup_{t\to\infty} \frac{1}{t} \log |Z(t)| \le -\lambda, \quad \text{a.s.}$$

and since $X(t) = \rho(t)Z(t)$ and $\lim_{t\to\infty} \log \rho(t)/t = \lambda$ a.s. we have that

$$\limsup_{t\to\infty} \frac{1}{t} \log |X(t)| \le 0, \quad \text{a.s.}$$

so that growth of solutions cannot be exponentially fast.

In the case when $a > 0$ and $X(0) > 0$, we can strengthen the result to

$$\lim_{t\to\infty} \frac{1}{t} \log |X(t)| = 0, \quad \text{a.s.},$$

so the subexponential dynamics u are inherited by X. This can be done by building a lower bound for $Z(t)$ by analogous comparison techniques to those outlined above (this works well, because Z is positive, and indeed, increasing), and prove that

$$\liminf_{t\to\infty} \frac{1}{t} \log Z(t) \ge -\lambda, \quad \text{a.s.},$$

which leads to

$$\liminf_{t\to\infty} \frac{1}{t} \log X(t) \ge 0, \quad \text{a.s.}$$

Combining this with the limsup, we get the claimed zero Liapunov exponent.

The argument in the case when $\lambda = 0$ is similar to that when $\lambda < 0$.

7 Proofs

Proof of Lemma 1 The proof up to (54) follows closely the line of Kato and McLeod (1971, Theorem 3(i)). By hypothesis, we rule out the case $a = 0$, since this reduces the equation to an ODE. With $a \neq 0$, there exists solutions $k \in \mathbb{C}$ to the equation $aq^k + b = 0$. Let k_0 be any solution of this equation. Then $\kappa = \Re(k_0)$ is uniquely defined and given by $\kappa = -\log(|b/a|)/\log(1/q)$. Let $c = \log q < 0$. Then $e^c = q$. Define $w : (-\infty, \infty) \to \mathbb{C}$ by $w(s) := e^{-sk_0} z(e^s)$ for $s \in \mathbb{R}$. Then

$$w(s + c) = e^{-sk_0} z(qe^s)/q^{k_0}, \tag{51}$$

and since z is differentiable on $(0, \infty)$, for all $s \in \mathbb{R}$, $w'(s)$ exists and is given by

$$w'(s) = e^s b(w(s) - w(s + c)) + e^{s-sk_0}\varphi(e^s) - k_0 w(s),$$

using (51) and $aq^{k_0} + b = 0$. Gathering together the $w(s)$ terms on the righthand side and treating the equation as a perturbed linear ODE, we are led to

$$\frac{d}{ds}\left(w(s)e^{k_0 s - be^s}\right) = -be^s e^{k_0 s - be^s} w(s+c) + e^{-be^s} e^s \varphi(e^s). \tag{52}$$

Next, for $n \in \mathbb{N}$, define $I_n = [s_0 - nc, s_0 - (n+1)c]$ (note $c < 0$), $\tau_n := s_0 - nc$ and $K_n := \sup_{s \in I_n} |w(s)|$. Now, let $s \in I_{n+1}$ and integrate (52) over $[s_0 - (n+1)c, s]$:

$$w(t)e^{k_0 t - be^t}\Big|_{t=\tau_{n+1}}^{s} = -b \int_{\tau_{n+1}}^{s} e^t e^{k_0 t - be^t} w(t+c)\,dt + \int_{\tau_{n+1}}^{s} e^{-be^t} e^t \varphi(e^t)\,dt.$$

Now, since $b < 0$, $-b = |b| > 0$ and $s \in I_{n+1}$, taking the complex modulus on both sides, and using the triangle inequality, we get

$$\begin{aligned} |w(s)||e^{k_0 s}|e^{|b|e^s} \le{} & |w(s_0 - (n+1)c)||e^{k_0(s_0-(n+1)c)}|e^{|b|e^{s_0-(n+1)c}} \\ & + |b| \int_{s_0-(n+1)c}^{s} e^t |e^{k_0 t}| e^{|b|e^t} |w(t+c)|\,dt + \int_{s_0-(n+1)c}^{s} e^{|b|e^t} e^t |\varphi(e^t)|\,dt. \end{aligned}$$

Since $\mathrm{Re}(k_0) = \kappa$, we get

$$\begin{aligned} |w(s)|e^{\kappa s + |b|e^s} \le{} & |w(-(n+1)c)|e^{\kappa(s_0-(n+1)c)} e^{|b|e^{s_0-(n+1)c}} \\ & + |b| \int_{s_0-(n+1)c}^{s} e^t e^{\kappa t + |b|e^t} |w(t+c)|\,dt + \int_{s_0-(n+1)c}^{s} e^{|b|e^t} e^t |\varphi(e^t)|\,dt. \end{aligned}$$

Since the terms in w are bounded by K_n, and the term in φ by ϵ_{n+1}, we get

$$\begin{aligned} |w(s)|e^{\kappa s + |b|e^s} \le{} & K_n \left(e^{\kappa(s_0-(n+1)c)} e^{|b|e^{s_0-(n+1)c}} + \int_{s_0-(n+1)c}^{s} |b|e^t e^{\kappa t + |b|e^t}\,dt \right) \\ & + |b| \int_{s_0-(n+1)c}^{s} e^{|b|e^t} e^t\,dt \cdot \frac{\epsilon_{n+1}}{|b|}. \end{aligned}$$

Now evaluate the last integral, and bound it by the antiderivative at the upper limit:

$$\begin{aligned} |w(s)|e^{\kappa s + |b|e^s} \le{} & K_n \left(e^{\kappa(s_0-(n+1)c)} e^{|b|e^{s_0-(n+1)c}} - \int_{s_0-(n+1)c}^{s} be^t e^{\kappa t + |b|e^t}\,dt \right) \\ & + \frac{\epsilon_{n+1}}{|b|} e^{|b|s}. \end{aligned} \tag{53}$$

Next, notice that we may take s_0 so large that $\kappa - k + |b|/2e^{s_0 - kc} > 0$ for all integers $k \ge 0$, which makes $t \mapsto \kappa t + |b|e^t$ monotone on each I_n, for any n. We focus on the integral remaining on the righthand side in (53). Integrating by parts gives

$$\int_{s_0-(n+1)c}^{s} be^t e^{\kappa t+|b|e^t}\,dt = \frac{be^t}{\kappa - be^t} e^{\kappa t+|b|e^t}\Big|_{s_0-(n+1)c}^{s} + \mathrm{O}\left(e^{\kappa s+|b|e^a}\int_{s_0-(n+1)c}^{s}\left|\frac{d}{ds}\frac{be^t}{\kappa - be^t}\right|dt\right),$$

where the big–O term is uniform in s, κ, s_0 and n, provided $s \in I_{n+1}$ and the stipulations on s_0 hold. The integral can therefore be further estimated according to

$$-\left(1-\frac{\kappa}{be^t}\right)^{-1} e^{\kappa t+|b|e^t}\Big|_{s_0-(n+1)c}^{s} + O\left(\kappa e^{-s_0+(n+1)c} e^{\kappa s+|b|e^s}\right) = -e^{\kappa t+|b|e^t}\Big|_{s_0-(n+1)c}^{s} + O\left(\kappa e^{-s_0+(n+1)c} e^{\kappa s+|b|e^s}\right).$$

Now, we return to (53) with this estimate to get

$$|w(s)|e^{\kappa s+|b|e^s} \le \frac{\epsilon_{n+1}}{|b|} e^{|b|s} + K_n\left(e^{\kappa(s_0-(n+1)c)} e^{|b|e^{s_0-(n+1)c}} + e^{\kappa t+|b|e^t}\Big|_{s_0-(n+1)c}^{s} + O\left(\kappa e^{-s_0+(n+1)c} e^{\kappa s+|b|e^s}\right)\right).$$

Thus for $s \in I_{n+1}$ we have $|w(s)| \le K_n\left\{1 + O\left(\kappa e^{-s_0+(n+1)c}\right)\right\} + \epsilon_{n+1}e^{-\kappa s}/|b|$. Taking the sup on both sides over I_{n+1}, and noting that $\kappa < 0$ yields

$$K_{n+1} \le K_n(1 + Ae^{nc}) + \frac{\epsilon_{n+1}}{|b|} e^{-\kappa(s_0-(n+2)c)} = K_n(1 + Ae^{(n+1)c}) + B'\epsilon_{n+1}e^{\kappa cn},$$

where $A > 0$ and $B' > 0$. Since $e^{-\kappa c} = \alpha \in (0, 1)$, with $\lambda_{n+1} = \epsilon_{n+1}/\alpha^{n+1}$ we have $K_{n+1} \le K_n(1 + Ae^{nc}) + B\lambda_{n+1}$ for $n \ge 0$. Thus for $n \ge 2$

$$K_n \le \prod_{j=0}^{n-1}(1 + Ae^{jc}) \cdot K_0 + B\sum_{j=1}^{n-1}\prod_{l=j}^{n-1}(1 + Ae^{lc}) \cdot \lambda_j + \lambda_n.$$

Let $p_n = \prod_{j=0}^{n-1}(1 + Ae^{jc})$ for $n \ge 1$. Then $p_n \to p_* \in (1, \infty)$ as $n \to \infty$ and so

$$K_n \le p_n K_0 + Bp_n \sum_{j=1}^{n} \frac{\lambda_j}{p_j}, \quad n \ge 2. \tag{54}$$

Thus, if $\sum_{n=1}^{\infty} \lambda_n < +\infty$, then $K_n \le K^*$ for all $n \ge 0$. On the other hand, if $\sum_{j=1}^{n} \lambda_j \to \infty$ as $n \to \infty$, then $K_n \le K^* \sum_{j=1}^{n} \lambda_j$ for all $n \ge 1$, by applying Toeplitz lemma to the sum on the righthand side, proving both parts of the lemma.

If λ_n is summable, $K_n \leq K^*$. Since $\kappa < 0$, $\sup_{t\in[e^{s_0-nc}, e^{s_0-nc}e^{-c}]} |z(t)| \leq K^*[e^{(s_0-nc)}]^\kappa$. Next, for all t sufficiently large there is a (unique) $n \in \mathbb{N}$ such that $e^{s_0-nc} \leq t < e^{s_0-nc}e^{-c}$. Since $\kappa < 0$, $t^\kappa \geq e^{\kappa(s_0-nc)}e^{-\kappa c} = \alpha e^{\kappa(s_0-nc)}$. Thus

$$|z(t)| \leq \sup_{t\in[e^{s_0-nc}, e^{s_0-nc}e^{-c}]} |z(t)| \leq K^*[e^{(s_0-nc)}]^\kappa \leq K^* \frac{1}{\alpha} t^\kappa,$$

as required. In the case λ_n is not summable, $K_n \leq K^* \sum_{j=1}^n \lambda_j$. Since $\kappa < 0$, we have

$$e^{-\kappa(s_0-nc)} \sup_{s\in[s_0-nc, s_0-(n+1)c]} |z(e^s)| \leq \sup_{s\in[s_0-nc, s_0-(n+1)c]]} e^{-\kappa s}|z(e^s)| = K_n,$$

so

$$\sup_{t\in[e^{s_0-nc}, e^{s_0-nc}e^{-c}]} |z(t)| \leq K_n \alpha^n \left(\frac{e^{-\kappa c}}{\alpha}\right)^n e^{\kappa s_0} = K_n \alpha^n e^{\kappa s_0} \leq K^{**} \alpha^n \sum_{j=1}^n \frac{\epsilon_j}{\alpha^j}.$$

as required. □

Proof of Theorem 2 Write $\varphi(t) = \mathrm{O}(\psi(t)t^{\kappa+1})$. We prove the following subcases:

(A) If $\psi \in \mathrm{RV}_\infty(-1)$, $\psi \in L^1(0,\infty)$, then $z(t) = \mathrm{O}(t^\kappa)$ as $t \to \infty$.
(B) If $\psi \in \mathrm{RV}_\infty(-1)$, $\psi \notin L^1(0,\infty)$, then $z(t) = \mathrm{O}\left(t^\kappa \int_0^t \psi(s)\,ds\right)$, $t \to \infty$.
(C) If $\psi \in \mathrm{RV}_\infty(\beta)$ for $\beta > -1$, then the estimate in (B) holds. Equivalently, if $\varphi = \mathrm{O}(\gamma)$ where $\gamma \in \mathrm{RV}_\infty(\eta)$ and $\eta > \kappa$, then $z(t) = \mathrm{O}(\gamma(t))$ as $t \to \infty$.
(D) If $\psi \in \mathrm{RV}_\infty(\beta)$ for $\beta < -1$, then the estimate in (A) holds. Equivalently, if $\varphi = \mathrm{O}(\gamma)$ where $\gamma \in \mathrm{RV}_\infty(\eta)$ and $\eta < \kappa$, then $z(t) = \mathrm{O}(t^\kappa)$ as $t \to \infty$.

Since $t \mapsto t^{\kappa+1}\psi(t)$ is regularly varying, and an estimate of the form $|\varphi(t)| \leq Kt^{\kappa+1}\psi(t)$ holds, by the uniform convergence theorem for regularly varying functions, $\epsilon_n = \sup_{s\in[e^{s_0-nc}, e^{s_0-nc}e^{-c}]} |\varphi(s)| \leq \bar{K}e^{(s_0-nc)(\kappa+1)}\psi(e^{s_0-nc})$. Since ψ is regularly varying, we get $\epsilon_n \leq K^*e^{-nc(\kappa+1)}\psi(e^{-nc})$, so $\lambda_n = \epsilon_n/\alpha^n \leq K^*\psi(e^{-nc})e^{-nc}$. By the uniform convergence theorem for regularly varying functions

$$\frac{\int_x^{xe^{-c}} \psi(s)\,ds}{\psi(x)x} = \int_1^{e^{-c}} \frac{\psi(\eta x)}{\psi(x)}\,d\eta \to \int_1^{e^{-c}} \eta^\beta\,d\eta =: k > 0, \quad x \to \infty,$$

and so we have that

$$e^{-nc}\psi(e^{-nc}) \sim \frac{1}{k}\int_{e^{-nc}}^{e^{-(n+1)c}} \psi(s)\,ds, \quad n \to \infty, \tag{55}$$

For part (A), ψ is integrable, so (55) shows that $(e^{-nc}\psi(e^{-nc}))$ is summable, and hence (λ_n) is summable. Hence by the fundamental lemma, $z(t) = \mathrm{O}(t^\kappa)$, $t \to \infty$.

For part (D), since $\beta < -1$, ψ is integrable, and by the same argument $z(t) = O(t^\kappa)$, $t \to \infty$. The second half of the claim follows by making the identification $\psi = \gamma/t^{\kappa+1}$, and since $\eta < \kappa$, $\psi \in \mathrm{RV}_\infty(\beta)$ with $\beta = \eta - \kappa - 1 < -1$.

For (B) and (C), ψ is not integrable by hypothesis. Using (55), we have

$$\sum_{j=1}^n \lambda_j \le K^* \sum_{j=1}^n \psi(e^{-jc})e^{-jc} \le K^{**} \sum_{j=1}^n \frac{1}{k} \int_{e^{-jc}}^{e^{-(j+1)c}} \psi(s)\,ds \le \bar{K} \int_1^{e^{-(n+1)c}} \psi(s)\,ds.$$

Now, since $p_n > 1$ and $p_n \to p^*$ as $n \to \infty$, by (54), we get

$$\sup_{t\in[e^{s_0-nc}, e^{s_0-nc}e^{-c}]} |z(t)| \le \alpha^n K_n \le \alpha^n \left(p_n K_0 + p_n \sum_{j=1}^n \lambda_j/p_j \right)$$

$$\le C_1\alpha^n + C_2\alpha^n \sum_{j=1}^n \lambda_j \le C_1\alpha^n + C_2\alpha^n \bar{K} \int_1^{e^{-(n+1)c}} \psi(s)\,ds.$$

Since the integral diverges, we have the consolidated estimate

$$\sup_{t\in[e^{s_0-nc}, e^{s_0-nc}e^{-c}]} |z(t)| \le C_3\alpha^n \int_1^{e^{-(n+1)c}} \psi(s)\,ds.$$

Finally, take $t \in [e^{s_0-nc}, e^{s_0-(n+1)c}]$. Then $e^{-nc}e^{-c} \le e^{-s_0}e^{-c}t$. Since ψ is regularly varying, and ψ is not integrable, $\int \psi$ is also regularly varying. Hence

$$\int_1^{e^{-(n+1)c}} \psi(s)\,ds \le \int_1^{e^{-c}e^{-s_0}t} \psi(s)\,ds \le C_4 \int_1^t \psi(s)\,ds.$$

On the other hand, $e^{-nc} \le e^{-s_0}t$, so $\alpha^n = e^{-nc\kappa} \ge e^{-\kappa s_0}t^\kappa$. Thus

$$|z(t)| \le \sup_{t\in[e^{s_0-nc}, e^{s_0-nc}e^{-c}]} |z(t)| \le C_3 \cdot e^{-\kappa s_0}t^\kappa \cdot C_4 \int_1^t \psi(s)\,ds,$$

as required. In part (B), and for the first part of the claim in (C), this suffices.

For the second claim in (C), we have $\varphi = O(\gamma)$, with $\gamma \in \mathrm{RV}_\infty(\eta)$ and $\eta > \kappa$. This allows us to take $\psi(t) = \gamma(t)/t^{\kappa+1} \in \mathrm{RV}_\infty(\beta)$ with $\beta = \eta - \kappa - 1 > -1$, so the hypotheses in the first part of the claim apply. Further, by the regular variation of ψ,

$$\int_0^t \psi(s)\,ds \sim \frac{1}{\beta+1} t\psi(t) = \frac{1}{\eta-\kappa}\frac{\gamma(t)}{t^\kappa}, \quad t \to \infty,$$

and so applying the first part of the claim gives $z(t) = O(\gamma(t))$, $t \to \infty$.

The conclusions of the different parts of the theorem can be combined. Parts (A) and (B) can be combined: let $\varphi = \mathrm{O}(\gamma)$ where $\gamma \in \mathrm{RV}_\infty(\kappa)$. Then $\psi(t) := \gamma(t)/t^{\kappa+1}$ for $t \geq 1$ is in $\mathrm{RV}_\infty(-1)$, and $\varphi(t) = \mathrm{O}(t^{\kappa+1}\psi(t))$. Then, in (A) and (B), we have

$$z(t) = \mathrm{O}\left(t^\kappa \int_1^t \psi(s)\,ds\right) = \mathrm{O}\left(t^\kappa \int_1^t \frac{\gamma(s)}{s^{\kappa+1}}\,ds\right), \quad t \to \infty.$$

For part (C), $\eta > \kappa$, so ψ is in $\mathrm{RV}_\infty(\eta - \kappa - 1)$. Since $\eta - \kappa - 1 > 0$, the function in the big O–estimate is asymptotic to $\gamma(t)/(\eta - \kappa)$ as $t \to \infty$, so by part (C), $z(t) = \mathrm{O}(\gamma(t)) = \mathrm{O}\left(t^\kappa \int_1^t \frac{\gamma(s)}{s^{\kappa+1}}\,ds\right)$ as $t \to \infty$, which agrees with (A) and (B). Finally, for (D), where $\eta < \kappa$, $\psi \in \mathrm{RV}_\infty(\eta - \kappa - 1)$. Since $\eta - \kappa - 1 < -1$, $\psi \in L^1(1,\infty)$, and so by (D)

$$z(t) = \mathrm{O}(t^\kappa) = \mathrm{O}\left(t^\kappa \int_1^t \frac{\gamma(s)}{s^{\kappa+1}}\,ds\right), \quad t \to \infty.$$

This allows the desired consolidation of the statement of the theorem. □

8 Conclusions

The paper considers perturbations of the asymptotically stable deterministic pantograph equation

$$u'(t) = bu(t) + au(qt)$$

where $a \neq 0$. As is well–known, all solutions of this equation tend to zero if and only if $b < 0$ and $|a| < |b|$. We thus presume these conditions hold, and ask what is the asymptotic behaviour of forced equations

$$x'(t) = bx(t) + ax(qt) + f(t),$$
$$dX(t) = bX(t) + aX(qt) + f(t)\,dt + \sigma(t)\,dB(t).$$

We can characterise, in terms of f and σ, necessary and sufficient conditions such that solutions are bounded or convergent to zero, assuming only that the underlying equation has stable solutions. These conditions depend on the time averages of f and σ over moving compact intervals, but do not depend on the parameters (a, b and q) of the underlying unperturbed equation. The conditions allow f and σ to behave very badly on a pointwise basis, while being satisfied by often–assumed stability conditions such as $f(t) \to 0$ or $f \in L^1(0,\infty)$. The conditions on the stochastic perturbation are more delicate; the a.s. stability or a.s. boundedness is captured by the convergence or divergence of an infinite sum which depends on a parameter. This condition is known to characterise the a.s. stability or a.s. boundedness of solutions of the stochastic ordinary differential equation

$$dY_0(t) = -Y_0(t)\,dt + \sigma(t)\,dB(t)$$

a result proven in an earlier work. The necessary and sufficient conditions are a little opaque, but a very good sufficient condition is that

$$\sigma_1^2(n)\log n \to 0, \quad n \to \infty; \tag{56}$$

where $\sigma_1^2(n) = \int_{n-1}^{n} \sigma^2(s)\,ds$; indeed, if σ_n^2 is decreasing, (56) is necessary, so once again we see that it is the average behaviour of σ^2 over moving compact intervals that determines the asymptotic behaviour.

A key observation in the entire analysis is that the stability of the deterministic functional differential equation is inherited from the ordinary differential equation

$$y' = -y + f$$

and the SFDE from $Y = Y_0 + y$, which solves a "doubly perturbed" SDE. It turns out Y tends to zero if and only if both Y_0 and y are, while Y is bounded if and only if Y_0 and y are.

As well as characterising boundedness and convergence, the approach enables us to determine very sharp conditions describing the rate of growth or decay of deterministic equations. If the solution is to grow, or has a growing running maximum, the rate of growth of its running maximum is that of y, and this can be characterised entirely through the time averages $\int_{t-\theta}^{t} f(s)\,ds$, which has the same asymptotic order of y. In terms of estimating the growth order, and characterising when certain growth orders of solutions occur, these results are unimprovable. Crucially, they show that relying on pointwise estimates on f may give very poor and misleading asymptotic information.

When solutions tend to zero, they do so slowly in most cases, since the unperturbed equation has solutions which tend to zero at the rate t^κ, where $\kappa < 0$ is known in terms of a, b and q. Existing analysis for general equations with unbounded delay, applied to the pantograph equation, gives estimates of $x(t) = O(f(t))$ when $f(t) = O(t^\beta)$ and $f'(t) = O(t^{\beta-1})$, and $\beta > \kappa$, and $x(t) = O(t^\kappa)$ when $f(t) = O(t^\beta)$ for $\beta < \kappa$. We show, in contrast, that when $y(t) = O(\gamma(t))$ and γ is regularly varying, then

$$x(t) = O\left(t^\kappa \int_1^t \frac{\gamma(s)}{s^{\kappa+1}}\,ds\right), \quad t \geq 0$$

which shows that neither pointwise nor derivative bounds are needed on f. Moreover, the estimate we establish is sharper in some cases than existing estimates, is never weaker than existing estimates, and is unimprovable in at least some situations (for equations in which $a > 0$).

The rate of growth of y plays a determining role the asymtotic behaviour when y dominates the solution of the unperturbed equation. For instance, if γ is regularly varying with index greater than κ, or γ is increasing, then $x(t) = o(\gamma(t))$ if and only if $y(t) = o(\gamma(t))$, $x(t) = O(\gamma(t))$ if and only if $y(t) = O(\gamma(t))$ and $\limsup_{t\to\infty} |x(t)|/\gamma(t) \in (0, \infty)$ if and only if $\limsup_{t\to\infty} |y(t)|/\gamma(t) \in (0, \infty)$.

We have sufficient conditions which describe when the rate of decay of the underlying pantograph equation is preserved. If $y(t) = O(\gamma(t))$, γ is regularly varying, and $t \mapsto \gamma(t)/t^{\kappa+1}$ is integrable, then $x(t) = O(t^\kappa)$ as $t \to \infty$. These conditions seem, in certain cases at least, unimprovable, since we have shown by means of general examples, that if the integrability condition is relaxed, the unperturbed rate of decay will not be preserved. Good necessary conditions prove for the moment, more elusive. We know that if $x(t) = O(t^\kappa)$, then $y(t) = O(t^\kappa)$; but taking the sufficient condition above into account, it seems that the maximal allowable rate of growth of y would be such that $t \mapsto |y(t)|/t^{\kappa+1} \in L^1(0, \infty)$, a condition satisfied in the case $y(t) = O(t^\kappa\{\log t \log\log t (\log\log\log t)^2\}^{-1})$, for instance. Thus, our best converse result in the case of a small perturbation is still suboptimal.

Although it might seem strange that we can characterise asymptotic bounds for "large" perturbations, but not "small" ones, we saw by means of an example that it is possible to arrange for the solution of the perturbed equation to decay at an arbitrarily fast rate, for a very carefully chosen and very rapidly decaying forcing term. Therefore, there is potentially great sensitivity in the rate of decay of solutions when there are small forcing terms, and this probably explains in part why it is hard to prove a very refined converse. As a first step though, our converse is not unacceptable: the above discussion suggests that our estimate on the critical y is only off by a factor which grows at around the (very slow) rate of $\log t$ as $t \to \infty$.

In this paper, we have not attempted to give quantitative asymptotic estimates for the solutions of stochastic equations. In part, this is because a characterisation of the rates of decay and growth of solutions of (26) is needed, and this task forms part of work in progress. However, once the task of characterising the asymptotic rates of growth and decay of Y is completed, the analysis in this paper (see Theorems 10 and 9) should be reusable with minor modifications, with Theorems 2 and 4 applicable to $Z = X - Y$, with the estimate on φ being read off from Y.

There are two other directions where the results of this paper could readily be developed. One is in the direction of "unstable" equations, which is to say equations where the condition (2) do not hold, and the unperturbed equation has unbounded solutions. The case when $b > 0$ should be more straightforward, since it that case, solutions of the unperturbed equation grow like e^{bt}, so one would hope that the solution of the perturbed equation would grow at the faster rate between e^{bt} and γ, where γ describes the rate of growth of y.

The case when $b < 0$, and $|a| > |b|$ is more interesting, since in that case, the unperturbed equation exhibits power law growth at rate t^κ; now $\kappa > 0$. In this situation, we expect analogues of Theorems 9 and 10 to hold. Revisiting the proof of the fundamental Lemma 1 reveals that the estimates hold up to (54), with perhaps some cosmetic changes needed earlier. Therefore it is probable that we will be able to estimate as in the proof of Theorem 2. As a result, we expect a theorem analogous to

Theorems 9 to hold if $y(t) = O(\gamma(t))$ and γ is regularly varying with index greater than κ (or γ is increasing faster than a regularly varying function), while a theorem analogous to Theorem 10 should hold if γ has index less than or equal to κ.

In Sect. 5, we commented that generalisations to equations with unbounded delay are possible, and we were able to sketch convergence and boundedness characterisation results directly. However, if growth bounds on solutions are required, we noted but that further work would be required, both on stochastic ordinary differential equations, as well as the analysis of Cermak in the deterministic case in order to sharpen estimates. Extensions of the convergence and boundedness results to the finite dimensional case are feasible too, but at the cost of stronger sufficient conditions being imposed on the structure of the unperturbed equation. We conjecture that a perturbation theorem which states that property (39) holds under the conditions (41) and (42) is true, but do not have immediate ideas how to proceed with it. The situation wherein there is multiplicative noise has been studied more extensively in the literature, but very refined results (i.e., results which classify the dynamics) appear to be lacking. However, by using the spirit of the approach taken here (decomposition of the solution into the solution of a stochastic ordinary differential equation, and the C^1 solution of a random functional differential equation), we have produced a result in the spirit of the very first results of Kato and McLeod: exponential asymptotic behaviour when the underlying non–delay equation is unstable, and subexponential asymptotic behaviour when the underlying non–delay equation is asymptotically stable. It is appealing that such progress is made for an equation with multiplicative noise by a multiplicative decomposition, since the bulk of this paper deals with additive noise, and is (we believe) successfully tackled by an additive decomposition of the solution. Further exploration of the multiplicative decomposition proposed here warrants a deeper investigation.

We finish with a remark that when perturbations are "small", the solution x of the perturbed pantograph equation is $O(t^\kappa)$. But it is well–known that solutions of the unperturbed equation exhibit log–periodic oscillation in the case that $a < 0$. With further careful analysis, it may be possible to show, for sufficiently small forcing terms, that the perturbed equation inherits this oscillatory asymptotic behaviour, even in the stochastic case. Such results would require developing analogues of Kato and McLeod's arguments. Since some of their analysis involves repeated differentiation of the equation, the double mollification ideas (or even higher order mollifications) sketched in Sect. 6.2 will be necessary if the results are not to require estimates of f and its derivatives in the deterministic case; indeed, such mollification would be necessary if one wished to prove more refined asymptotic results for stochastic equations, as suggested in Sect. 6.2.

Acknowledgements EL was supported by Science Foundation Ireland (16/IA/4443) while a postgraduate student at Dublin City University. JA is supported by the RSE Saltire Facilitation Network on Stochastic Differential Equations: Theory, Numerics and Applications (RSE1832).
The authors wish to thank the anonymous referees for their very careful reading of the manuscript, and for prompting many interesting questions concerning generalisations of the original submission. This has resulted in a much more comprehensive paper, prompting among other things the treatment of general unbounded delay, multidimensional equations, and multiplicative noise.

We also wish to thank Prof. Elena Braverman for the opportunity to present this work at the session on "Delay and stochastic differential equations with in Modelling in Finance, Life Sciences, and engineering" at ICIAM2023, and for the chance to submit it to this special volume.
Finally, the first author also wishes to acknowledge the love and support of his late wife, Huizhong Appleby–Wu, with whom he discussed early results in this paper; She also was in Tokyo in August 2023, and gave advice on the presentation of our talk at the ICIAM meeting. HA–W died suddenly and unexpectedly, at the age of 41, in January 2024, and before this paper was submitted for publication. May she rest in peace.

Competing Interests The authors have no conflicts of interest to declare that are relevant to the content of this chapter.

References

Appleby, J.A.D.: Decay and growth rates of solutions of scalar stochastic delay differential equations with unbounded delay and state dependent noise. Stoch. Dyn. **5**(2), 133–148 (2005)

Appleby, J.A.D., Buckwar, E.: A constructive comparison technique for determining the asymptotic behaviour of linear functional differential equations with unbounded delay. Differ. Equ. Dyn. Syst. **18**(3), 271–301 (2010)

Appleby, J.A.D., Buckwar, E.: Sufficient conditions for polynomial asymptotic behaviour of the stochastic pantograph equation. Electron. J. Qual. Theory Differ. Eqn. Proc. 10th Coll. Qual. Theory Differ. Eqn. (2), 1–32 (2016)

Appleby, J.A.D., Cheng, J., Rodkina, A.: Characterisation of the asymptotic behaviour of scalar linear differential equations with respect to a fading stochastic perturbation. Discrete Contin. Dyn. Syst. 79–90 (2011)

Appleby, J.A.D., Lawless, E.: Mean square asymptotic stability characterisation of perturbed linear stochastic functional differential equations. Appl. Numer. Math. **200**, 80–109 (2024)

Appleby, J.A.D., Lawless, E.: Solution space characterisation of perturbed linear functional and integrodifferential Volterra convolution equations: Cesàro limits, p. 21 (2024). arXiv:2407.08416

Appleby, J.A.D., Lawless, E: Weighted L_∞ Asymptotic Characterisation of Perturbed Autonomous Linear Ordinary and Stochastic Differential Equations: Part I—ODEs, p. 40 (2024)

Baker, C.T.H., Buckwar, E.: Continuous θ-methods for the stochastic pantograph equation. Electron. Trans. Numer. Anal. **11**, 131–151 (2000)

Baker, C.T.H., Buckwar, E.: Exponential stability in p-th mean of solutions, and of convergent Euler-type solutions, of stochastic delay differential equations. J. Comput. Appl. Math. **184**, 404–427 (2005)

Bingham, N.H., Goldie, C.M., Teugels, J.L.: Regular Variation. Encyclopedia of Mathematics and its Applications. Cambridge University Press (1991)

Carr, J., Dyson, J.: The functional differential equation $y'(x) = ay(\lambda x) + by(x)$. Proc. R. Soc. Edinb. A **74**, 165–174 (1974/1975)

Carr, J., Dyson, J.: The matrix functional differential equation $y'(x) = Ay(\lambda x) + By(x)$. Proc. R. Soc. Edinb. A **75**, 5–22 (1975/1976)

Čermák, J.: A change of variables in the asymptotic theory of differential equations with unbounded delay. J. Comput. Appl. Math. **143**(1), 81–93 (2002)

Čermák, J., Nechvátal, L.: On stability of linear differential equations with commensurate delayed arguments. Appl. Math. Lett. **125**, Paper No. 107750, 8 (2022)

Caraballo, T., Mchiri, L., Mohsen, B., Rhaima, M.: p-th moment exponential stability of neutral stochastic pantograph differential equations with Markovian switching. Commun. Nonlinear Sci. Numer. Simul. **102**, Paper No. 105916, 19 (2021)

Chan, T., Williams, D.: An "excursion" approach to an annealing problem. Math. Proc. Cambridge Philos. Soc. **105**(1), 169–176 (1989)

Diblík, J.: Asymptotic representation of solutions of the equation $y'(t) = \beta(t)[y(t) - y(t - \tau(t))]$. J. Math. Anal. Appl. **217**, 200–215 (1998)

Diblík, J., Růžičková, M.: Convergence of the solutions of the equation $y'(t) = \beta(t)[y(t - \delta) - y(t - \tau)]$ in the critical case. J. Math. Anal. Appl. **331**, 1361–1370 (2007)

Fan, Z., Liu, M., Cao, W.: Existence and uniqueness of the solutions and convergence of semi-implicit Euler methods for stochastic pantograph equations. J. Math. Anal. Appl. **325**(2), 1142–1159 (2007)

Fox, L., Mayers, D.F., Ockendon, J.R., Tayler, A.B.: On a functional differential equation. J. Inst. Math. Appl. **8**, 271–307 (1971)

Gripenberg, G., Londen, S.-O., Staffans, O.: Volterra Integral and Functional Equations. Encyclopedia of Mathematics and its Applications. Cambridge University Press (1990)

Guo, P., Li, C.-J.: Almost sure stability with general decay rate of exact and numerical solutions for stochastic pantograph differential equations. Numer. Algorithms **80**(4), 1391–1411 (2019)

Haddock, J.R., Krisztin, T.: On the rate of decay of solutions of functional differential equations with infinite delay. Nonlinear Anal. **10**(8), 727–742 (1986)

Hartman, P., Wintner, A.: Asymptotic integration of ordinary nonlinear differential equations. Am. J. Math. **77**, 692–724 (1955)

Hartman, P.: Ordinary Differential Equations, 2nd edn. SIAM, Philadelphia (2002)

Hu, Y., Wu, F., Huang, C.: Stochastic stability of a class of unbounded delay neutral stochastic differential equations with general decay rate. Int. J. Syst. Sci. **43**(2), 308–318 (2012)

Huang, Y., Yan, Z., Zheng, H.: Stochastic H_∞ filtering for pantograph systems with state dependent noise. In: 2012 IEEE International Conference on Automation and Logistics (ICAL), 5–17 August 2012, Zhengzhou, pp. 601–605 (2012)

Iserles, A.: On the generalized pantograph functional-differential equation. Eur. J. Appl. Math. **4**, 1–38 (1993)

Jiang, F., Yang, H., Wang, S.: Asymptotic stability of stochastic pantograph differential equations with Markovian switching. J. Nonlinear Anal. Optim. **1**(1), 9–16 (2010)

Karatzas, I., Shreve, S.E.: Brownian Motion and Stochastic Calculus. Graduate Texts in Mathematics, vol. 113. Springer, New York (1991)

Kato, T., McLeod, J.B.: The functional-differential equation $y'(x) = ay(\lambda x) + by(x)$. Bull. Am. Math. Soc. **77**, 891–937 (1971)

Krisztin, T.: Asymptotic estimation for functional differential equations via Lyapunov functions. Colloquia Math. Soc. János Bolyai **I**(109), 365–376 (1988)

Lachouri, A., Samei, M.E., Ardjouni, A.: Existence and stability analysis for a class of fractional pantograph q-difference equations with nonlocal boundary conditions. Bound. Value Probl. Paper No. 2, 20 (2023)

Liu, Y.: On functional differential equations with proportional delays. Ph.D. Thesis, Department of Mathematics, Cambridge University, Cambridge

Liu, X., Chen, T.: Robust μ-stability for uncertain stochastic neural networks with unbounded time-varying delays. Phys. A. **387**(12), 2952–2962 (2008)

Makay, G., Terjéki, J.: On the asymptotic behavior of the pantograph equations. Electron. J. Qual. Theory Differ. Eqn. **1998**(2), 1–12 (1998)

Mao, W., Hu, L., Mao, X.: The asymptotic stability of hybrid stochastic systems with pantograph delay and non-Gaussian Lévy noise. J. Franklin Inst. **357**(2), 1174–1198 (2020)

Mao, X.: Almost sure polynomial stability for a class of stochastic differential equations. Q. J. Math. Oxford Ser. **43**(2), 171, 339–348 (1992)

Mao, X.: Polynomial stability for perturbed stochastic differential equations with respect to semimartingales. Stoch. Process. Appl. **41**(1), 101–116 (1992)

Mao, X.: Stochastic Differential Equations and Their Applications. Horwood Publishing Limited, Chichester (2008)

Meng, X., Hu, S., Wu, P.: Pathwise estimation of stochastic differential equations with unbounded delay and its application to stochastic pantograph equations. Acta Appl. Math. **113**(2), 231–246 (2011)

Meng, X., Yin, B.: On the general decay stability of stochastic differential equations with unbounded delay. J. Korean Math. Soc. **49**(3), 515–536 (2012)

Milosevic, M.: Existence, uniqueness, almost sure polynomial stability of solution to a class of highly nonlinear pantograph stochastic differential equations and the Euler-Maruyama approximation. Appl. Math. Comput. **237**, 672–685 (2014)

Ockendon, J.R., Tayler, A.B.: The dynamics of a current collection system for an electric locomotive. Proc. R. Soc. A **322**, 447–468 (1971)

Pavlovic, G., Jankovic, S.: Razumikhin-type theorems on general decay stability of stochastic functional differential equations with infinite delay. J. Comput. Appl. Math. **236**(7), 1679–1690 (2012)

Pituk, M.: The Hartman-Wintner theorem for functional differential equations. J. Differ. Eqn. **155**(1), 1–16 (1999)

Pituk, M.: A Perron type theorem for functional differential equations. J. Math. Anal. Appl. **316**(1), 24–41 (2006)

Shiryaev, A.N.: Probability, 2nd edn. Springer, Berlin (1996)

Siva Ranjani, A., Suvinthra, M., Balachandran, K.: Large deviations for stochastic pantograph integrodifferential equation. Filomat, **37**(20), 6751–6766 (2023)

Strauss, A., Yorke, J.A.: Perturbation theorems for ordinary differential equations. J. Differ. Eqn. **3**, 15–30 (1967)

Strauss, A., Yorke, J.A.: On asymptotically autonomous differential equations. Math. Syst. Theory **1**, 175–182 (1967)

Wu, F., Hu, S.: Razumikhin-type theorems on general decay stability and robustness for stochastic functional differential equations. Int. J. Robust Nonlinear Control **22**(7), 763–777 (2012)

Wu, H., Hu, J., Yuan, C.: Stability of numerical solution to pantograph stochastic functional differential equations. Appl. Math. Comput. **431**, Paper No. 127326, 13 (2022)

Yan, Z., Huang, Y.: Robust H_∞ filter design for Itô stochastic pantograph systems. Math. Probl. Eng. Art. ID 747890, 8 (2013)

Yang, H., Jiang, F., Jiang, Y.: Robust stability of stochastic pantograph differential equations Markovian switching. In: International Conference on Computer Application and System Modelling (ICCASM), vol. 5, pp. 541–545. October 22–24, 2010, Shanxi, Taiyuan (2010)

MIX
Papier aus verantwortungsvollen Quellen
Paper from responsible sources
FSC® C105338
FSC
www.fsc.org

If you have any concerns about our products,
you can contact us on
ProductSafety@springernature.com

In case Publisher is established outside the EU,
the EU authorized representative is:
Springer Nature Customer Service Center GmbH
Europaplatz 3, 69115 Heidelberg, Germany

Printed by Libri Plureos GmbH
in Hamburg, Germany